Strange But True
SCIENCE

150 ANSWERS to questions you never thought to ask

Publications International, Ltd.

Cover images: All from Shutterstock.com except upper left by Cynthia Fliege

Interior images: Shutterstock.com

Contributing writers: Angelique Anacleto, Brett Ballantini, Diane Lanzillotta Bobis, Joshua D. Boeringa, Shelley Bueché, Michelle Burton, Steve Cameron, Matt Clark, Anthony G. Craine, Dan Dalton, Paul Forrester, Shanna Freeman, Chuck Giamatta, Ed Grabianowski, Jack Greer, Tom Harris, Vickey Kalambakal, Brett Kyle, Noah Liberman, Letty Livingston, Alex Nechas, Jessica Royer Ocken, Thad Plumley, ArLynn Leiber Presser, Pat Sherman, William Wagner, Carrie Williford

ISBN: 978-1-4508-9324-4

Manufactured in U.S.A.

8 7 6 5 4 3 2 1

Table of Contents

Teleportation, meteor strikes, Vomit Comets, and a few matters of gravity.

The puzzles of planet Earth, from the origins of dew to the legend of the green flash.

The mysteries of the eternal fruitcake and the holey Swiss cheese revealed.

Big Blue, preserving corpses, time zones...and we make a good brown paper bag too.

All about the human animal and what happens to it after eating turkey, whiffing smelling salts, or getting a metal plate stuck in its head.

Everyday Mysteries

Does running through the rain keep you drier than walking? Here are the answers to the daily mysteries that keep you pacing the floor at night.

The Body Strange

If you really think you want the details behind spontaneous combustion, bodies donated to science, or life after being beheaded...read on.

Odd Things about Animals

You've smelled the elephant and heard the owl call you names, but what about the headless chicken, the raining fish, and the radioactive roach?

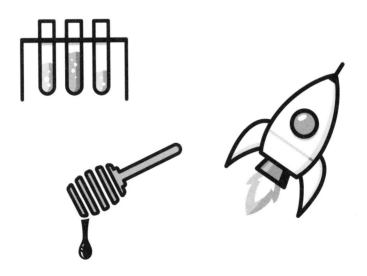

Introduction

Science is strange. You always suspected that something weird lurked inside those college textbooks you used to keep your dorm window open. Well, let us confirm that suspicion for you. We've carefully sifted through some truly eccentric studies and data sets to assemble a collection of highly entertaining explanations for life's most maddening mysteries. Namely, why is the bird poop on your windshield white? How did they get the salt inside the peanut shell? What makes your voice go squeaky when you inhale helium?

That's right. You'll be able to explain to everyone at the cocktail party or convention exactly how scratch-and-sniff works. Your vast knowledge of weird and provocative topics will win you innumerable bar bets. You'll finally be able to prove to your spouse where the dirtiest place in the house is. You'll be the smart aleck your parents always claimed you were.

Strange But True Science! is an authoritative narrative on myriad topics, ranging from human and animal biology to earth science and astronomy. Our writers demystify urban legends and inform on everything from the chemically routine to the cosmically curious. Page through this offering and get smart. How far do you have to dive underwater to escape gunfire? Read about it before you need to find out at that cocktail party.

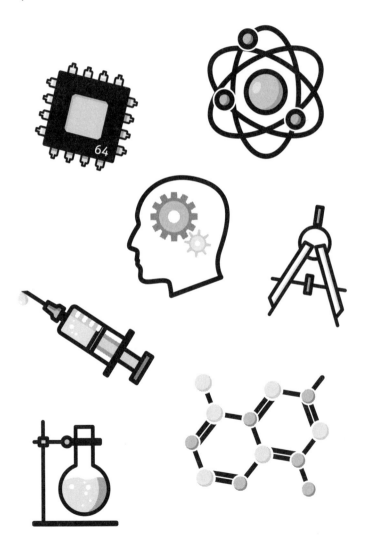

Space Oddities

Do weird things happen to an astronaut's body in outer space?

They sure do. Playing zero-gravity paddleball and cruising around on the surface of the moon may look like good, clean fun, but space travel is no picnic. Astronauts' bodies endure some crazy changes in the celestial firmament, and it takes awhile for them to recover once they're back on solid ground.

So what's the problem? In a word, weightlessness. And the most immediate consequence of the zero-G lifestyle is something that astronauts call space adaptation syndrome. It occurs because the structure of the inner ear that gives you your sense of balance is acclimated to the constant force of gravity; when that force disappears, your inner ear tells you that you're perpetually falling forward. This typically causes nausea, vomiting, dizziness, and disorientation.

There are other, more serious consequences from leaving

Earth, and they have to do with gravity's effect on the rest of the body. Here on Earth, we spend our entire lives within the planet's gravitational field. Even when we're not trying to leap over a puddle or pull off a Superman dunk, the force of gravity is constantly compressing our bodies—and, importantly, our bodies are fighting back.

To understand how your body is forever fighting gravity, just think of your circulatory system. The force of gravity tries to pull your blood supply down into your lower extremities; your heart meanwhile works hard against gravity to keep the blood flowing into your upper body, too. If you were to leave Earth's gravitational field, your heart would be working too hard at forcing the blood upwards, causing your face to swell and leading to nasal congestion and bulging eyes. But eventually, your body would adjust to its new environment; your heart would pump less intensely and your blood pressure would lower.

And without the force of gravity, your muscles atrophy—they get smaller and weaker. Hardest hit are postural muscles like your hamstrings and back muscles, the ones that fight against gravity to keep you standing tall. But all of your muscles begin to wither, as it requires less effort to make any movement. Even your heart is affected, in part because of your lowered blood pressure. As a perk, your intervertebral discs—essentially shock absorbers for the spine—expand, making you two to three inches taller. But even this height enhancement can be painful.

The biggest problem is the effect of weightlessness on the bones. In order to stay strong enough to meet the demands of daily living, your bones constantly regenerate themselves in accordance with the level of strain that they experience. For example, if you lift weights regularly, your bones will grow stronger and more calcium-fortified.

In space, however, the reduced level of stress causes bones to weaken and lose their mass. Studies on astronauts working in Skylab in the 1970s showed a 0.3 percent loss of bone mass during each month of weightlessness.

Due to all of these physical changes, astronauts are in pretty bad shape when they get back home. Their lowered blood pressure can lead to fainting. (To alleviate this problem, they sometimes wear special suits that compress the legs and feet, forcing more blood to the torso and head.) Their sense of balance is out of whack for about a week, making it hard for them to remain steady on their feet. And they're very weak because of the muscle degeneration; it can take months to fully regain lost muscle mass. Bone regeneration can take years, and extended missions—like a three-year Mars trip—would cause permanent bone damage.

It's a lot of wear and tear for the opportunity to hit a golf ball on the moon. Rocketing into space doesn't sound quite as cool anymore, does it?

Are we going to be hit by a meteor?

We already have been, and we will be again.

NASA estimates that about once every hundred years, a rocky asteroid or an iron meteorite substantial enough in size to cause tidal waves hits Earth's surface. About once every few hundred thousand years, an object strikes that is large enough to cause a global catastrophe.

NASA's Near Earth Objects program scans the skies and observes comets and asteroids that could potentially enter Earth's neighborhood. It has been keeping close tabs on an asteroid called Apophis, a.k.a. MN2004. According to NASA, on April 13, 2029, Apophis will be close enough to Earth that it will be visible to the naked eye. At one time, the odds were estimated to be as great as one in three hundred that Apophis would hit Earth. However, NASA has now ruled out a collision, which is a good thing because the asteroid would have hit Earth with the force of an 880-megaton explosion (more than fifty thousand times the power of the atomic bomb dropped on Hiroshima, Japan, in 1945).

Perhaps the best-known meteor hit occurred fifty thousand years ago, when an iron meteorite collided with what is now northern Arizona with a force estimated to be two thousand times greater than the bomb dropped on Hiroshima. Now named the Meteor Crater, the twelve-thousand-meters-wide crater is a popular tourist attraction.

A direct meteor hit isn't even necessary to cause signifi-cant damage. On June 30, 1908, what many believe was a small asteroid exploded high in the air near the Tunguska River in Russia. Taking into consideration the topography of the area, the health of the adjoining forest, and some new models concerning the dynamics of the explosion, scientists now believe that the force of the explosion was about three to five megatons. Trees were knocked down for hundreds of square miles.

NASA hopes to provide a few years' warning if there is a meteor approaching that could cause a global catastrophe. The organization anticipates that our existing technology would allow us to, among other things, set off nuclear fusion weapons near an object in order to deflect its trajectory. Or we can simply hope that Bruce Willis will save us, just like he did in the 1998 movie *Armageddon*.

Is the Great Wall of China the only manmade object visible from space?

The problem here lies with the phrasing of the question, specifically the "visible from space" bit. Now, if you were in a spaceship in far-off space, with a nice, powerful telescope—sure, you'd be able to see the Great Wall pretty easily. But most people mean "visible with the naked eye."

That leads us to the second thing that's ambiguous about the question. "Space" is a tricky term to define. A lot of people swap "space" with "the moon" when asking this question. So, is the Great Wall of China the only manmade object that's visible with the naked eye from the moon? Not a chance. No manmade object is visible from the moon. At an altitude of just a few thousand miles above Earth's surface, no manmade objects are visible. (We say "a few thousand miles" because nobody's really put a finger on the magic point at which it all disappears—astronauts have far more important things to do.) The moon is about 240,000 miles away. You can barely see the continents from up there, let alone the Great Wall. If you went out to Mars, Earth would appear to be just a fairly bright star in the sky.

There's hope, though. "Space" is a lot closer than people think. It begins about sixty miles above Earth's surface. Until you get somewhere between one hundred fifty and three hundred miles above the surface (vague for the same reasons as before), the Great Wall is visible with the naked eye—if the weather and lighting are exactly right. You'll still have to squint a bit, though, since it's a similar color to the landscape around it.

But before you start throwing this fact around trying to win bar bets, know this: From that height, lots of manmade objects are visible, including highways, railways, and sailing ships. Cities are, of course, quite easy to distinguish from the countryside, especially at night when there are millions of lights on. From the lowest areas of space, you can even see some individual buildings and airplane contrails.

So after making a short story long, here's the answer: Nope, it ain't true.

What would happen if Earth stopped spinning?

You know when you slam on the brakes in your car and the CDs and soda cans go flying? Now imagine slamming on the brakes when you're going 1,100 miles per hour, the planet's rotational speed at the equator. The instant that Earth stopped spinning, its atmosphere and inhabitants—along with soil, plants, buildings, oceans, and everything else that isn't firmly attached to the rocky foundation of the planet's crust—would keep on going at 1,100 miles per hour. The face of the planet would be wiped clean.

Let's say you were up in the Space Shuttle and missed all the planet-wiping excitement. What would life be like when you got back to now-still Earth? The good news is that there would be no change in gravity, which means that you wouldn't fall off the planet and the atmosphere wouldn't go away. But you would notice plenty of other differences. First of all, the cycles of day and night as we know them would no longer exist. Wherever you were, it would be light for about six months and then dark for about six months. As a result, one side of the planet would be icy cold and the other side would be extremely hot.

The planet's overall wind patterns would change significantly, too. Major wind patterns are caused by the sun heating the planet unevenly. The sun's rays hit the equator directly and the North Pole and South Pole at an angle, which means that the area around the equator gets much hotter than the mass around the poles. This heat gradient continually drives warmer air toward the poles and cooler air toward the equator, which establishes a basic global wind pattern.

But the spinning motion of the planet complicates this basic northerly and southerly airflow, creating smaller wind systems called convection cells in each hemisphere and leading to prevailing easterly and westerly winds. These systems interact to generate the weather that dictates the climate around the globe. If Earth didn't spin, we wouldn't see the same complex weather patterns. Warm air would simply rise at the equator and rush to the poles, and cold winds would move the opposite way.

Finally, a non-spinning Earth would stop generating a magnetic field. Yes, compasses would be useless, but there would be a much bigger problem: Earth would no longer possess the magnetic field's protection against cosmic rays.

The radiation from the sun and other stars would damage your DNA, leading to severe health problems like cancer. But the extreme heat or cold and total lack of animal and plant life would kill you well before the nasty radiation kicked in.

Don't fret, though. There is virtually no chance that any of this could happen. For Earth's rotational speed to change radically, it would need to collide with an asteroid the likes of which we've never seen. Even if that happened, it's extremely unlikely that the collision would stop the planet from spinning altogether—it would probably just slow it down. In any case, we would see something that big well in advance, which would give Bruce Willis enough time to go and blow it up.

If the moon causes tides, why does high tide happen twice a day?

We've all been taught that the moon causes tides because of its gravitational pull. This is somewhat accurate, but our teachers weren't telling us the whole truth.

Let's start simply and then add the more complicated bits. The moon revolves around Earth, right? Wrong: Earth and the moon revolve around each other. More specifically, they revolve around a specific point, called the barycentre, where the moon and Earth balance each other out. This point is inside Earth, but it's closer to the surface than the core.

So Earth and the moon revolve around this point. Got it so far? Good. The moon exerts more pull on the side of Earth to which it's closest, which explains why you get a true high tide on only one side of the planet at a time. The moon pulls the ocean so that the water bulges a bit, and the nearby surfers are in for some sweet waves.

Because Earth is spinning around this barycentre, it is affected by a centrifugal force that points away from the moon. (Actually, "centrifugal force" in this case is something of a misnomer. Scientists like to talk about "rotating reference frames" and other silliness, but if you substitute the term "centrifugal force," it offers basically the same result and is simpler to understand. Scientists seem to be fond of making things as complicated as possible.) The moon exerts less gravitational pull on the far side of Earth, and centrifugal force takes charge, creating another bulge there. Two bulges at opposite ends mean two high tides a day.

So that about wraps up this question, neatly and tide-ily. (Sorry.)

Why did Space Shuttle astronauts wear parachutes?

NASA devised an escape system for Space Shuttle missions after the 1986 Challenger disaster, in which seven astronauts died when a rocket booster exploded shortly after liftoff. The parachutes were part of a coordinated plan that offered them a chance to bail out if problems arose during launch or landing.

For obvious reasons, jumping from the shuttle is impossible while its rockets are firing. But there are scenarios in which escape would be an option. One would be after the rockets finish firing but before the shuttle reaches space. Another would be if the rockets fail after launch and the astronauts face a dangerous emergency landing in the ocean.

How would an escape work? First, the crew would guide the shuttle to an altitude of about twenty-five thousand to thirty thousand feet—just lower than the altitude reached by commercial airline flights—and jump from the craft through a side hatch.

To avoid hitting a wing or an engine pod during their escape, the astronauts would extend a twelve-foot pole from the side of the shuttle, hook themselves to it, slide down, and jump from there. NASA's space suits are designed to work automatically during an escape. The parachute opens at fourteen thousand feet, and when the suit detects impact with water, the parachute detaches.

Astronauts have other gizmos up their sleeves (and pant legs) that help in an emergency. When water is detected, the suit automatically deploys a life preserver. Also contained within the suit is a life raft, complete with a bailing cup to remove water that sloshes into it. Once safely afloat, the astronaut can pull a set of flares from one leg pocket and an emergency radio from the other. The suit, which is designed to keep the astronaut alive for twenty-four hours, is pressurized, thermal, and even comes equipped with a supply of drinking water.

How long is a day on Mars?

The Martian solar day lasts about twenty-four hours and forty minutes. That's not much longer than a day on Earth, but it would give Earthlings who might eventually colonize the red planet a substantial advantage. Think about how much more you could accomplish with an extra forty minutes per day. It would amount to about an extra twenty hours per month.

Those colonists would need to adjust their calendar as well as their clock. A year on Mars is about 687 Earth days. This is because Mars has a much longer path of orbit around the sun. Earth zips around the sun almost twice before Mars completes one full circuit.

Because of its extended path of orbit, Mars experiences four seasons that last twice as long as those on Earth. Spring and summer are almost two hundred days each; fall and winter are about one hundred and fifty days each. Martian farmers would have a lot of time to plant and harvest, but they'd also have to conserve that harvest through a much longer, much harsher winter.

Everything lasts longer on Mars than on Earth. For those colonists, the Martian work week would be an extra three hours and twenty minutes, and the weekend would be extended by an hour and twenty minutes. Summer vacation would be long, but the school year would be longer still.

A blessing or a curse? Depends on what you're doing with your extra forty minutes per day.

Is there such a thing as a blue moon?

> *Blue moon,*
> *You saw me standing alone,*
> *Without a dream in my heart,*
> *Without a love of my own.*
> —From the song "Blue Moon"

According to the many performers who have recorded this Rodgers and Hart classic—including Elvis Presley, Frank Sinatra, and Bob Dylan—there most certainly is such a thing as a blue moon, and it acts as a celestial matchmaker for the lovelorn. Of course, not everyone is as sappy as the aforementioned singers. When most people mention a blue moon, they are referring to an event that is highly unusual. As our lovelorn crooners might say, "I have a date once in a blue moon."

The phrase "blue moon" dates back to 1528. It first appeared in a work by William Barlow, an English bishop, the wonderfully titled *Treatyse of the Buryall of the Masse*. "Yf they saye the mone is belewe," Barlow wrote, "we must beleve that it is true." (Trust us; he's saying something about a blue moon here.) After Barlow's usage, which no doubt confused as many readers as it edified, the term came to represent anything absurd or impossible.

It was only later that "blue moon" connoted something unusual. Most etymologists trace this usage to the wildly popular 1819 edition of the *Maine Farmer's Almanac*, which suggested that when any season experiences four full moons (instead of the usual three), the fourth full moon was to be referred to as "blue."

As is often the case with these things, somehow the *Maine Farmer's Almanac*'s suggestion was misinterpreted—researchers blame the incompetent editors of a 1946 issue of *Sky & Telescope* magazine—to mean a second new full moon in a single month. Consequently, in present-day astronomy, that second new full moon is referred to as a "blue" moon. This frequency, ironically, isn't all that unusual, at least as astronomical events go: once every two and a half years.

As for whether the moon really can appear blue, the answer is yes. After massive forest fires swept through western Canada in 1950, for example, much of eastern North America was treated to a bluish moon in the night sky. However, events such as this occur, well, once in a blue moon.

How do astronauts go to the bathroom?

Weightlessness sure seems fun. You see those astronauts effortlessly floating around, mugging for the camera, and magically spinning their pens in midair. But what you don't get to see is what happens when nature calls. You can be sure that as much as they enjoy swimming through the air like waterless fish, there's one place on Earth where all astronauts thank their lucky stars for gravity: the bathroom.

On the Space Shuttle, the astronaut sat on a commode with a hole in it, not unlike a normal toilet—except for the restraints that fit over the feet and thighs to prevent his or her body from floating away. Suction took the place of gravity, so the seat was cushioned, allowing the astronaut's posterior to form an airtight seal around the hole. If everything was situated properly, the solid waste went down the main hole: A separate tube with a funnel on the end took care of the liquids. With so much going on, relaxing with a newspaper was not really an option.

Today's astronauts have it easy compared to their forebears on the Apollo missions (1961–1975). When an Apollo astronaut had to go number two, he attached a specially designed plastic bag to his rear end. The bag had an adhesive flange at its opening to ensure a proper seal.

But if you think that this procedure couldn't have been any more undignified, consider this: There was no privacy. The astronauts would usually carry on with their duties while they were, you know, doing their duty. In the words of Apollo astronaut Rusty Schweickart, "You just float around for a while doing things with a bag on your butt." With no gravity and no suction, getting the feces to separate from the body was, generally, an hour-long process. It began with removing the bag—very carefully—and ended with lots and lots of wiping.

Where does all this stuff go? Fecal material is dried, compressed, and stored until the ship returns to Earth. (Some scientists believe that manned missions to Mars will require waste to be recycled and used for food. If you were hoping to sign up for one of those flights, you may want to think twice before dropping your application in the mail.) Urine, on the other hand, is expelled into space. The memory of this procedure caused Schweickart to wax darn-near poetic, calling a urine dump at sunset, "one of the most beautiful sights" he saw in space.

"As the stuff comes out and hits the exit nozzle," Schweickart went on, "it instantly flashes into ten million little ice crystals, which go out almost in a hemisphere. The stuff goes in every direction, all radially out from the spacecraft at relatively high velocity. It's surprising, and it's an incredible stream of...just a spray of sparklers almost. It's really a spectacular sight."

And you thought stars looked cool.

Will there be an elevator to outer space someday?

Rockets are for suckers. Scientists at NASA looking for a cheaper way to build space stations and satellites have hit upon an idea that sounds ludicrous but that they swear is feasible: an elevator that reaches from Earth's surface to outer space.

"How could that be possible?" you ask. "What have these scientists been smoking?" The answer is nanotubes. (That's not what they've been smoking—it's the technology that could make a space elevator possible.) Discovered in 1991, nanotubes are cylindrical carbon molecules that make steel look like a ninety-eight-pound weakling. A space elevator's main component would be a sixty-thousand-odd-mile nano-tube ribbon, measuring about as thin as a sheet of paper and about three feet wide.

It gets weirder. That ribbon would require a counterweight up at the top to keep it in place. The counterweight, hooked to the nanotube ribbon, would be an asteroid pulled into Earth's orbit or a satellite. Once secured, the ribbon would have moving platforms attached to it. Each platform would be powered by solar-energy-reflecting lasers and could carry several thousand tons of cargo up to the top. The trip would take about a week. Transporting materials to outer space in this fashion would supposedly reduce the cost of, say, putting a satellite into orbit from about ten thousand dollars a pound to about one hundred dollars a pound.

The base of the elevator would be a platform situated in the eastern Pacific Ocean, near the equator, safe from hurricanes and many miles clear of commercial airline routes. The base would be mobile so that the whole thing could be moved out of the path of potentially damaging space junk orbiting Earth. Although there are a lot of theoretical kinks to work out, the more optimistic of the scientists who have hatched this scheme believe the whole thing could be a reality within a couple of decades. Think about that the next time you step into an elevator and chug on up to the third floor.

What happens to all the stuff we launch into space and don't bring back?

Space trash creates a major traffic hazard. If you think it's nerve-wracking when you have to swerve around a huge pothole as you cruise down the highway, just imagine how it would feel if you were hundreds of miles above the surface of Earth, where the stakes couldn't be higher. That's the situation that the crew of the International Space Station (ISS) faced in 2008 when it had to perform evasive maneuvers to avoid debris from a Russian satellite.

And that was just one piece of orbital trash—all in all, there are tens of millions of junky objects that are larger than a millimeter and are in orbit. If you don't find this worrisome, imagine the little buggers zipping along at up to seventeen thousand miles per hour. Worse, these bits of flotsam and jetsam constantly crash into each other and shatter into even more pieces.

The junk largely comes from satellites that explode or disintegrate; it also includes the upper stages of launch vehicles, burnt-out rocket casings, old payloads and experiments, bolts, wire clusters, slag and dust from solid rocket motors, batteries, droplets of leftover fuel and high-pressure fluids, and even a space suit. (No, there wasn't an astronaut who came home naked–the suit was packed with batteries and sensors and was set adrift in 2006 so that scientists could find out how quickly a spacesuit deteriorates in the intense conditions of space.)

So who's responsible for all this orbiting garbage? The two biggest offenders are Russia–including the former Soviet Union–and the United States. Other litterers include China, France, Japan, India, Portugal, Egypt, and Chile. Each of the last three countries has launched one satellite during the past twenty years.

Most of the junk orbits Earth at between 525 and 930 miles from the surface. The ISS operates a little closer to Earth–it maintains an altitude of about 250 miles–so it doesn't see the worst of it. Still, the ISS's emergency maneuver in 2008 was a sign that the situation is getting worse. Houston, we have a problem.

NASA and other agencies use radar to track the junk and are studying ways to get rid of it for good. Ideas such as shooting at objects with lasers or attaching tethers to some pieces to force them back to Earth have been discarded because of cost considerations and the potential danger to people on the ground. Until an answer is found, NASA practices constant vigilance, monitoring the junk and watching for collisions with working satellites and vehicles as they careen through space. Hazardous driving conditions, it seems, extend well beyond Earth's atmosphere.

Why aren't there southern lights?

There are. The southern lights are called the "aurora australis," and according to those who've seen them (including famed explorer Captain James Cook, who named the lights in 1773), they are just as bright and alluring as the aurora borealis in the north. We don't hear about them because the viewing area—around the geomagnetic South Pole—is mostly unpopulated.

Northern or southern, the lights are the result of solar storms that emit high-energy particles. These particles travel from the sun as a solar wind until they encounter and interact with the earth's magnetic field. They then energize oxygen atoms in the upper atmosphere, causing light emissions that can appear to us as an arc, a curtain, or a green glow. If these oxygen atoms get really excited, they turn red.

There are other atoms in the ionosphere, and they produce different colors when they're titillated by those solar winds. Neutral nitrogen will produce pink lights, and nitrogen radicals glow blue and violet.

Usually, the lights are visible only in latitudes between ninety degrees (at the poles) and thirty degrees. In the north, that large swath includes most of Europe, Asia (excluding India, except for its northernmost tip, and southern countries such as Myanmar, Thailand, and Cambodia), the United States, and Canada. In the south, though, only the southern-most tips of Australia and Africa and the countries of Chile, Argentina, and Uruguay in South America are within that zone.

So in reality the question is this: If a light shines in the south and there is no one there to see it, does it still dazzle?

Where did the moon come from?

Before Neil and Buzz had a chance to poke around up there, astronomers had three competing theories to explain the origin of the moon, based strictly on their observations of it from a great distance.

The co-accretion, or condensation, theory stated that the moon formed out of swirling space dust and gas at the same time as Earth, 4.5 billion years ago. The fission theory, meanwhile, speculated that the newly formed Earth was spinning so fast that it shed a bunch of debris that clumped together to form the moon. And the capture theory held that the moon was a wandering asteroid that meandered past Earth and got snagged by its gravitational pull.

Everyone hoped that hands-on analysis of the moon would clear things up. But the rocks that were collected by astronauts from the lunar surface didn't validate any of these explanations—on the contrary, each theory was cast into further doubt.

Moon rock turned out to be similar to Earth rock, which made the wandering asteroid idea seem unlikely. But oddly enough, moon rock is completely dry, while the minerals of Earth contain significant amounts of moisture. This discovery worked against the fission theory, since spun-off Earth debris would likely retain some water content. The fission theory was further undermined when the samples' densities suggested that the moon once had an entirely molten surface, which wouldn't have been possible if the debris had simply peeled off of Earth.

Finally, seismic readings taken by the astronauts showed that the moon is much less dense than Earth and has a much smaller inner core, if it has a core at all. This observation called the co-accretion theory into question. Why? Presumably, two planetary bodies that formed in the same spot at the same time would have more similarities in their structures than the seismic readings indicated.

The ensuing head-scratching sparked a fourth, more radical idea that is called the giant-impact, or giant-impactor, theory. In this scenario, the moon is the result of a Mars-size rock (a leftover from the solar system's birth) that slammed into Earth about 4.45 billion years ago. This collision sent huge chunks of Earth and the impactor into orbit. Some debris flew off into outer space, some debris fell back to Earth, and over the course of time, the rest of the hot rock in orbit clumped together to form the moon.

This is the leading theory today because it best fits the evidence. First, it's consistent with the composition of moon rock: The intense heat of the impact could have vaporized any water that was in the flying debris, and it could have given the moon its once-molten surface. The theory also explains why the moon may not have a core: It was formed mainly from Earth's crust material, so it has something similar to that consistency throughout. The many giant craters on the moon also support the notion that there were a lot of giant flying rocks zipping around the solar system in the old days.

All things considered, it appears as if Earth was lucky to lose only a moon-size chunk.

How close are we to teleporting, like they do in *Star Trek?*

Closer than you think, but don't squander those frequent-flyer miles just yet. There's a reason why Captain Kirk is on TV late at night shilling for a cheap-airfare Web site and not hawking BeamMeToHawaiiScotty.com. For the foreseeable future, jet travel is still the way to go.

If, however, you're a photon and need to travel a few feet in a big hurry, teleportation is a viable option.

Photons are subatomic particles that make up beams of light. In 2002, physicists at the Australian National University were able to disassemble a beam of laser light at the sub-atomic level and make it reappear about three feet away. There have been advances since, including an experiment in which Austrian researchers teleported a beam of light across the Danube River in Vienna via a fiber-optic cable—the first instance of teleportation taking place outside of a laboratory.

These experiments are a far cry from dematerializing on your spaceship and materializing on the surface of a strange planet to make out with an alien who, despite her blue skin, is still pretty hot. But this research demonstrates that it is possible to transport matter in a way that bypasses space—just don't expect teleportation of significant amounts of matter to happen until scientists clear a long list of hurdles, which will take many years.

Teleportation essentially scans and dematerializes an object, turning its subatomic particles into data. The data is transferred to another location and used to recreate the object. This is not unlike the way your computer downloads a file from another computer miles away. But your body consists of trillions upon trillions of atoms, and no computer today could be relied on to crunch numbers powerfully enough to transport and precisely recreate you.

As is the case with many technological advances, the most vexing, long-lasting obstacle probably won't involve creation of the technology, but moral and ethical issues surrounding its use.

Teleportation destroys an object and recreates a facsimile somewhere else. If that object is a person, does the destruction constitute murder? And if you believe that a person has a soul, is teleportation capable of recreating a person's soul within the physical body it recreates? And should we someday cross that final frontier, if BeamMeToHawaiiScotty.com becomes a reality, do you believe that William Shatner should star in the TV commercial?

Why does the moon look bigger when it's near the horizon?

If this one has you stumped, don't fret: It's flummoxed brilliant minds for thousands of years. Aristotle attempted an explanation around 350 BC, and today's scientists still don't know for sure what's going on. Great thinkers have, however, ruled out several possible explanations.

First, the moon is not closer to Earth when it's at the horizon. In fact, it's closer when it's directly overhead.

Second, your eye does not physically detect that the moon is bigger when it's near the horizon. The moon creates a .15-millimeter image on the retina, no matter where it is. You can test this yourself: Next time you see a big moon looming low behind the trees, hold a pencil at arm's length and note the relative size of the moon and the eraser. Then wait a few hours and try it again when the moon is higher in the sky. You'll see that the moon is exactly the same size relative to the eraser. The .15-millimeter phenomenon rules out atmospheric distortion as an explanation for the moon's apparent change in size.

Third, a moon on the horizon doesn't look larger just because we're comparing it to trees, buildings, and the like. Airline pilots experience the same big-moon illusion when none of these visual cues are present. Also, consider the fact that when the moon is higher in the sky and we look at it through the same trees or with the same buildings in the foreground, it doesn't look as large as it does when it's on the horizon.

What's going on? Scientists quibble over the details, but the common opinion is that the "moon illusion" must be the result of the brain automatically interpreting visual information based on its own unconscious expectations. We instinctively take distance information into account when deciding how large something is. When you see far-away building, for example, you interpret it as big because you factor in the visual effect of distance.

But this phenomenon confuses us when we attempt to visually compute the size of the moon. According to the most popular theory, this is because we naturally perceive the sky as a flattened dome when, in reality, it's a spherical hemisphere. This perception might be based on our understanding that the ground is relatively flat. As a result, we compute distance differently, depending on whether something is at the horizon or directly overhead.

According to this flattened-dome theory, when the moon is near the horizon, we have a fairly accurate sense of its distance and size. But when the moon is overhead, we unconsciously make an inaccurate estimate of its distance. As a result of this error, we automatically estimate its size incorrectly.

In other words, based on a faulty understanding of the

shape of the sky, the brain perceives reality incorrectly and interprets the moon as being smaller when it's overhead than when it's on the horizon. That's right—your brain is tricking you. So what are you going to believe—science or your lying eyes?

Why are planets round?

We all know about gravity. When something is really heavy (the size of a planet, for example), it pulls other objects toward it. That's why we don't float off Earth—it's holding us down. Technically, we're also pulling Earth toward us, but because we're so small and Earth is so large, we don't really affect it much.

Now, everything on a planet is being pulled toward its heavy center because of the planet's gravity, and everything is drawn as close to the center as possible. The only way for everything on the surface to be equally close to the center is for the planet to be round; each point on the surface of a sphere is the same distance from its center.

If a planet were, say, a cube, its corners would be farther from the center than everything else. Because of gravity, these corners would collapse inward to get closer to the center. After this collapse and the planet's subsequent reformation, it would end up being a good old sphere.

But like its human inhabitants, Earth is not perfect. It bulges slightly at its middle (the equator) because it is spinning. Its shape, then, is somewhat wonky; Earth's equator is more than a dozen miles farther from the planet's center than are the North and South poles. The technical term for Earth's shape is "oblate spheroid." Most planets are oblate spheroids. Saturn is the most noticeably wonky—its equator is 10 percent fatter than its polar diameter. Throw that at your friends the next time there's a lull in the conversation, and watch them bow to your genius.

Why is Pluto not a planet anymore?

Poor Pluto. It was welcomed into the exclusive club of planets in 1930 after being discovered by American astronomer Clyde Tombaugh, but then was unceremoniously booted out on August 24, 2006, by the International Astronomical Union (IAU). Because Pluto's orbit around the sun takes approximately 247.9 Earth years, it didn't even get to celebrate its first anniversary of being a planet.

Pluto didn't change. What did?

The word and original definition of "planet" are derived from the Greek *asteres planetai,* which means "wandering stars." Planets are known as wanderers because they appear to move against the relatively fixed background of the stars, which are much more distant. Five planets (Mercury, Venus,

Mars, Jupiter, and Saturn) are visible to the naked eye from Earth and were known to people in ancient times.

Three other planets would not have been discovered without modern advancements. Uranus was discovered by William Herschel in 1781, using a telescope; Johann Galle discovered Neptune in 1846, using sophisticated mathematical predictions; and Pluto was discovered when the astronomer Tombaugh laboriously flipped through photographic plates of regions of the sky captured at different dates that showed the special wandering that denotes a planet.

Continued improvements in telescopes and mathematical modeling, among other advancements, have helped astronomers find all sorts of things in a solar system that previously seemed somewhat vacant. Classifying all of these objects—especially with respect to what is a planet and what's not—has proven to be tricky. On August 24, 2006, the IAU established a cut-and-dried definition of a planet. To be a planet, an object must orbit the sun, have enough mass so that it is nearly round, and dominate the area around its orbit.

This turn of events was bad news for Pluto because it doesn't meet the third part of the definition—it doesn't dominate its "neighborhood." For one thing, planets are supposed to be much larger than their moons, and Pluto's moon, Charon, is about half its size. Second, a planet is supposed to clear the neighborhood around its orbit—meaning it should, in the words of *National Geographic News,* "sweep up aster-oids, comets, and other debris"—and Pluto isn't particularly effective at doing that.

The IAU didn't completely diss Pluto. It now is classified as a dwarf planet, meaning that it orbits the sun, is round, and isn't a satellite of any other object, despite not clearing its orbit. There are dozens of known dwarf planets, and scien-tists expect the number to grow rapidly because so many objects fit the criteria.

NASA also has given Pluto its due (though it did so before it was declassified as a planet). On January 19, 2006, NASA launched an unmanned probe called New Horizons that is bound for Pluto. After traveling three billion miles, the probe is expected to enter the Pluto system in the summer of 2015. It carries some of the ashes of Tombaugh, who no doubt would have been horrified to learn that his grand discovery has been reduced to a dwarf.

What is the Vomit Comet?

Actually, it's an airplane, any one of several owned by NASA and used over the past fifty years to train astronauts and conduct experiments in a zero-gravity environment.

The plane simulates the absence of gravity by flying in a series of parabolas—arcs that resemble the path of an especially gut-wrenching roller coaster. When the Vomit Comet descends toward the earth, its passengers experience weightlessness for the twenty to twenty-five seconds it takes to reach the bottom of the parabola. Then the plane flies back up to repeat the maneuver, beginning a new dive from an altitude of over thirty thousand feet.

Being weightless and buoyant might bring on nausea all by itself, but when the plane arcs, dips, and ascends again, the riders feel about twice as heavy as usual. The wild ride induces many of its otherwise steely-stomached passengers to vomit—hence, the name. (The plane is also called, by those with a greater sense of propriety, the Weightless Wonder.)

NASA has used the Vomit Comet to train astronauts for the Mercury, Gemini, Apollo, Skylab, Space Shuttle, and Space Station programs. The first Vomit Comets, which were unveiled in 1959 as part of the Mercury program, were C-131 Samaritans. A series of KC-135A Stratotankers came next. The most famous of these, the NASA 930, was retired in 1995 after twenty-two years of service as NASA's primary reduced-gravity research plane. This is the aircraft that was used to film the scenes of space weightlessness in the 1995 movie *Apollo 13.* It is now on public display at Ellington Field, near Johnson Space Center in Houston.

After the 930 was put out to pasture, another KC-135A took over, the NASA 931, which was retired in 2004. The 931 flew

34,757 parabolas, generating some 285 gallons of vomit. Yes, the engineers at NASA measured the barf. Since 2005, a C-9—the military version of the DC-9 aircraft produced by McDonnell Douglas—has been used to give astronauts a taste of weightlessness . . . and of bile.

Why do stars shine?

Because they're really, really hot. Super-duper hot.

What? You want a better answer than that? Well, okay then. A star forms when large balls of gas in space gravitate toward each other. All the gas begins to contract into one massive ball of gas, and this contraction creates a lot of heat.

Once the star gets hot enough, it's time for something called nuclear fusion to begin. A star like our sun is largely made up of hydrogen. When that hydrogen is pushed into a dense, hot little mass, the individual atoms of hydrogen collide and fuse, forming heavier atoms, like helium. At the same time, the colliding atoms release energy, which makes the gas even hotter. This forms a protostar, which is the beginning of a star.

Eventually, this energy seeps toward the outer bits of the mass, and the whole thing becomes one of the big flaming balls of gas that we've come to know and love. The energy doesn't stop there, though. It radiates outward as heat and light, and it's this energy radiation that we see as shining stars from down here on Earth. The sun is far enough away from Earth that it takes eight minutes for its light to reach us, but the radiation can burn our skin and give us cancer if we're not careful. So it's pretty powerful stuff.

At some point, the star uses up all available atoms and expands into a red giant before imploding into a small, dense white dwarf that slowly fades out. The sun is due to do this in about five billion years and, in the process, will pretty much obliterate Earth. Scientists predict that the sun's outer surface will expand, making Earth part of the sun's atmosphere.

Now, if you were wondering why stars shine only at night, it's because the sun's light is so bright during the day that it blocks light from other stars.

The Not-So-Natural World

What's a green flash?

The logical answer would be a comic book superhero—a cross between the Green Lantern and the Flash. But that isn't it. A green flash comes from nature, not the mind of a geeky writer. It's a phenomenon that occurs at sunset or sunrise, during which part of the sun seems to change in color to green or emerald. The term "flash" is used to describe the change because it is visible for only a second or two.

Green flashes are so rarely seen that they have reached an almost mythological status. Some say they don't really exist—that they're just a mirage; others insist they do exist, but only in remote parts of the world. There's a mountain of misinformation about green flashes, dating back at least to the science fiction pioneer Jules Verne. In his 1882 novel *Le Rayon-Vert*, Verne wrote that a person who was "fortunate enough once to behold [a green flash] is enabled to see closely into his own heart and to read the thoughts of others." A green flash is remarkable, yes . . . but not that remarkable.

A green flash is the result of three optical phenomena: refraction near the horizon, scattering, and absorption. Refraction is the bending of a light wave as it travels through another medium. Scattering occurs when a light wave travels through particles whose diameter is no more than one-tenth the length of the light wave. Absorption occurs when a light wave reaches a material whose electrons are vibrating at the same frequency as one or more of the colors of light.

At sunset, the image of the sun that you can see is slightly above the actual position of the sun. This is caused by refraction, which separates the solar light into wavelengths, or colors. Although the atmosphere barely absorbs yellow light, even a little bit of absorption can make a big difference when the sun is near the horizon. The blue light is scattered away. So, what you have is the ray of red light "setting" at a very particular moment and no longer reaching your eye, the ray of yellow light being absorbed and no longer reaching your eye, and the ray of blue light being scattered about the atmosphere and no longer reaching your eye. The result is a momentary ray of green light–a green flash.

If you want to see a green flash, we have some tips. First and foremost, be patient and accept that you may never see one. However, if you know when and from where to look, and under what conditions, you might become one of the lucky observers of a green flash.

Find a place from which you have an unobstructed view of

the horizon. Mountains are the best vantage point. Beaches are second-best because you can use the ocean line as your guide. Other high places, such as an in-flight airplane or a tall building, will also do. You need to be in an area where the sky is cloudless and the air is clean; if the air is dusty, smoggy, or hazy, the green wavelengths won't be transmitted. You'll also want to have binoculars, because a green flash is very small.

Finally, be smart. Remember when your mom would tell you not to stare at the sun? Smart lady. At sunset, don't look at the sun until it is nearly down; at sunrise, start looking just as the sun seems ready to peek over the horizon. Staring at that big ball of fire at the wrong time can permanently bleach the red-sensitive photopigment in your eyes, forever distorting your color perception. Happy hunting.

How far away is the horizon?

This is a question that can only be answered by doing a bunch of math—the kind of math that many of us were happy to leave behind in college. You need to take into account the radius of the earth, of course. Square roots make an appearance, as do a cosine and a few subscripts and superscripts. Feel a headache coming on?

Even before you do the math, you have to decide what distance you're after. Is it the distance of the straight line from your eyeball to the horizon, or is it the distance you'd have to walk from where you're standing to get to the point that you see on the horizon? Believe it or not, this makes a difference, because the curvature of the earth makes the walking distance farther than the straight-line distance. And the height of the person doing the seeing has a significant impact on the answer, so that has to be factored in, as well. Now our mathematical equation is turning into one of those accursed word problems.

For the sake of simplicity, let's just say we want the straight-line distance. To a six-foot-tall man who is standing in a rowboat on the ocean, the horizon is about three miles away—until he comes to his senses and sits down to avoid capsizing the boat. When he's seated, his view to the horizon is reduced significantly, down to about one mile.

There. That wasn't so bad, was it?

Does lightning strike the same place twice?

"Lightning never strikes the same place twice." Although this popular adage seems nearly as old as lightning itself, it's about as accurate as your average weatherman's seven-day forecast. The truth is, lightning can—and often does—strike the same place twice.

To understand why this belief is an old wives' tale, we need a quick refresher course on how lightning works. As Ben Franklin taught us, lightning is pure electricity. (Electricity is a result of the interplay between positive and negative charges.) During a thunderstorm, powerful winds create massive collisions between particles of ice and water within a cloud; these encounters result in a negatively charged electrical field. When this field becomes strong enough—during a violent thunderstorm—another electrical field, this one positively charged, forms on the ground.

These negative and positive charges want to come together, but like lovers in a Shakespearean tragedy, they need to overcome the resistance of the parental atmosphere. Eventually, the attraction grows too strong and causes an invisible channel—known as a "stepped leader"—to form in the air. As the channel reaches toward the ground, the electrical field on the earth creates its own channels and attempts to connect with the stepped leader. Once these two channels connect, electricity flows from the cloud to the ground. That's lightning.

Lightning is an amazing phenomenon. The average bolt is about fifty thousand degrees Fahrenheit, or about ten times the temperature of the sun's surface. During a typical thunderstorm, nearly thirty thousand lightning bolts are created. The National Oceanic and Atmospheric Administration estimates that more than twenty-five million bolts of lightning strike the earth each year.

Given that huge number, it's hard to believe that lightning doesn't strike the same place twice. In fact, it does–especially when the places in question are tall buildings, which can be struck dozens of times a year. According to the National Lightning Safety Institute, the Empire State Building is hit an average of twenty-three times a year.

But tall buildings aren't the only objects that attract multiple lightning strikes. Consider park ranger Roy Cleveland Sullivan. For most of his career, Sullivan roamed the hills of Virginia's Shenandoah National Park, watching for poachers, assisting hikers, checking on campers–and being struck by lightning.

From 1942 to 1977, Lightnin' Roy was struck by lightning seven times. His eyebrows were torched off, the nail on one of his big toes was blown off, his hair was set aflame, and he suffered various burns all over his body. Sullivan ultimately committed suicide. Who could blame him?

How much rain does it take to make a rain forest?

It takes eighty inches of rain per year to make a rain forest, but the scientists who categorize these things aren't picky. There should be no feelings of inadequacy among forests whose drops per annum don't quite make the cut; if a wooded area has a rate of precipitation that comes close to the eighty-inch mark, it will gladly be taken into the fold.

Rain falls about ninety days per year in a rain forest. As much as 50 percent of this precipitation evaporates, meaning that rain forests recycle their water supply. In non-rain forest areas, water evaporates and is transported (via clouds) to different regions. In a rain forest, however, the unique climate and weather patterns often cause the precipitation to fall over the same area from which it evaporated.

A rain forest is comprised of evergreen trees, either broadleaf or coniferous, and other types of intense vegetation—these regions collectively contain more than two-thirds of the plant species on the planet. There are two types of rain forests: tropical and temperate. Tropical rain forests are located near the equator; temperate rain forests crop up near oceanic coastlines, particularly where mountain ranges focus rainfall on a particular region.

Rain forests can be found on every continent except Antarctica. The largest tropical rain forest is the Amazon in South America; the largest temperate rain forest is in the Pacific Northwest, stretching from northern California all the way to Alaska.

At one time, rain forests covered as much as 14 percent of the earth, but that number is now down to about 6 percent. Scientists estimate that an acre and a half of rain forest—the equivalent of a little more than a football field—is lost every second. The trees are taken for lumber, and the land is tilled for farming. At that rate, scientists estimate, rain forests will disappear completely within the next forty years—and it will take a lot more than eighty inches of rain per year to bring them back.

Do rocks grow?

Some do, but you'd have to stare at them an awfully long time to notice the difference. Why? These rocks grow at a rate of about one millimeter every million years. Oh, and don't forget to hold your breath while you're watching and waiting, because they're under the ocean.

Known as iron-manganese crusts, these rocks grow on the surfaces of undersea mountains. But they aren't alive—they don't reproduce like we do. Instead, they slowly and steadily collect chemical elements from seawater. An estimated two-hundred billion tons of iron-manganese crusts sit on the floors of the world's oceans. Some of these rocks contain high concentrations of metals such as cobalt, nickel, and platinum, making them a potentially lucrative target for mining.

Scientists find iron-manganese crusts fascinating, in the way that only scientists can find something fascinating. Because these rocks have been growing slowly over millions of years, their chemical compositions hold some valuable secrets about changes in the chemistry and circulation of the oceans over time.

Examining the chemical makeup of these rocks helps scientists understand how the planet's geologic processes work and gives them clues about what sort of impact humans have had on the planet. These clues aren't exactly the stuff of an Agatha Christie detective novel, but don't try telling that to the men and women in lab coats.

Is global warming causing more hurricanes?

For the sake of answering this question, we'll presume the existence of global warming. (Among the general public, and even among those in the scientific community, a heated debate exists over whether humans are harming the atmosphere with our carbon dioxide pollution.) So, assuming global warming is in effect, what sorts of changes can we expect in the weather? Specifically, do the rising temperatures cause more hurricanes?

A 2007 release issued by the Intergovernmental Panel on Climate Change (IPCC) stated that while there has been an increase in hurricanes since 1995, a clear pattern has yet to emerge. The yearly hurricane average did not change much throughout most of the twentieth century—in North America, there were about five hurricanes a year; for the years between 1997 and 2006, the annual average was eight. There are scientists who are prepared to pin the increase on global warming; others think those scientists are crying wolf.

At the very least, it can be said with a degree of certainty that higher global temperatures create more hospitable conditions for hurricanes. In order for a hurricane to form, ocean waters must be at least eighty degrees Fahrenheit. When this temperature is reached, the atmosphere above the water becomes unstable—a fertile situation for tropical storms. It stands to reason that if Earth experiences an overall increase in temperature, ocean waters will reach this temperature more often. That's the speculation, anyway.

A less controversial topic involves hurricane intensity, which the IPCC says has probably worsened as a result of human activity. "More likely than not," the panel says—though these aren't the most definitive words ever spoken. The IPCC report predicts an increase in wind speeds and overall destructive capacities of hurricanes.

What can we glean from this information? There may or may not be more hurricanes because of increased levels of carbon dioxide in the atmosphere, but the eight or so we do get every year will knock our socks off.

What causes air turbulence?

For even those most comfortable with flying, a sufficiently bumpy patch can lead to firmly gripped armrests and white knuckles. For anyone with a fear of flying, turbulence is the stuff of nightmares. We've been exposed to more than enough mental images, thanks to movies and television, to make it easy to envision the plane taking a sudden nosedive.

Pilots are usually quick to reassure their passengers, but they neglect to give out the information that might bring down a nervous passenger's blood pressure and put color back in his or her knuckles. Understanding turbulence might be a step toward getting over our fear of these invisible speed bumps.

Turbulence is caused by air currents moving in unpredictable ways. Airplanes achieve flight by manipulating air above and below the wings in such a way that more air flows under the wing than over, creating more air pressure under the wing than over, and thus giving the craft the ability—at the proper speed—to leave the ground. Essentially, airplanes are riding on air currents. So, when the current shifts unpredictably, the pressure around the wing changes, resulting in the turbulence you feel in the cabin.

Air currents might be moving because of a difference in temperature—warm air rises, while cool air settles. Or they might be moving over a mountain and shifting the surrounding air as they follow the jutting face of the earth. An airliner also might experience turbulence when crossing the wake of another jet or while passing currents created by violent weather patterns. If an airliner crosses a jet stream, which is a relatively narrow and fast-moving current of air caused by Earth's rotation, it will always experience turbulence (flying with a jet stream, on the other hand, is a smooth ride). Jet streams cut across the United States anywhere above twenty thousand feet, and they influence the movement of storms and other weather patterns.

Turbulence is separated into six levels of severity: Light Turbulence, Light Chop, Moderate Turbulence, Moderate Chop, Severe Turbulence, and Extreme Turbulence. The word "turbulence" here indicates a change in altitude, and as the levels get higher, the altitude changes become more pronounced. The word "chop" indicates bumpiness, without a noticeable change in altitude, similar to taking a truck through a field or down an unpaved forest trail.

In the instance of Extreme Turbulence, the aircraft is impossible to control for a period of time. The craft may suffer structural damage, and this can lead to the plane falling out of the sky. Don't lose your resolve, though—this kind of thing is rare.

If you can unclench your jaw and release your grip on the armrest, pat your frightened neighbor on the arm and explain what's happening. It might relieve his or her stress, too.

Why does a seashell sound like the ocean?

Is that big spiral conch you picked up during last year's trip to Hawaii still whispering sweet nothings in your ear? Well, that isn't the roar of the blue Pacific you hear—it's nothing more than the barrage of ambient noise around you.

Ah, science can be so harshly unsentimental sometimes! Seashells don't really create any sound all by themselves. Inside, they're a labyrinth of hollow areas and hard, curved surfaces that happen to be really good reflectors of racket.

When you hold a seashell up to your ear, that shell is actually capturing and amplifying all the little noises occurring around you. These noises are usually so hushed that you don't even hear them unless you're paying very close attention. However, when they begin bouncing off the cavity of a shell, the echoes resonate more loudly into your ear. And what do you know? They sound a lot like ocean waves rolling up to shore.

It doesn't matter how far away you are from the sea, or even if you have a seashell. You can re-create the same "ocean sound" effect by simply cupping your hand, or a coffee mug, over your ear. Just be sure that mug is empty—or you'll really hear a splash.

What are those vapors rising from the road on a hot day?

Next time you see a wiggly puddle of vapor on the road, think of that guy in the movie who is lost in the desert and spots a lake on the distant horizon. Desperate for a life-saving drink, he stumbles and crawls but never reaches the lake. Why? Because there is no lake, just like there is no wiggly puddle on the road ahead. What you both have seen is an inferior mirage.

Sunlight makes the road, as well as the area directly above the pavement, hotter than the prevailing air temperature. This layer of hot air that hovers inches above the pavement refracts light that passes through it—in other words, the light gets bent. It's as if a mirror had been placed on the road: The bent image that you see is the reflection of light coming from the sky. The same thing can happen just above the ground in a hot desert.

What is often reflected by this low layer of hot air is the light of a blue sky. In the desert, this resembles a lake; on the road, it can resemble puddled water or maybe oil. Sometimes, you might even see the reflection of a distant car.

On a boring drive, this phenomenon can be a pleasant distraction. And unlike the crawling desert guy, you can deal with your thirst by reaching for the cool drink that's in your cup holder.

Why are most plants green?

Maybe they're envious of our ability to walk over to the sink and get a drink of water.

While that theory is certainly compelling, plants aren't green with envy. The green comes from a pigment called chlorophyll. Pigments are substances that absorb certain wavelengths of light and reflect others. In other words, pigments determine color—you see the wavelength of light that the pigment reflects. A plant is green because the chlorophyll in it is really good at absorbing red and blue light but lousy at absorbing green light.

You find a heaping helping of chlorophyll in plants because chlorophyll's job is to absorb sunlight for use in photosynthesis, which is the process of converting sunlight and carbon dioxide into food (carbohydrates) that plants need to survive. So, since a plant wouldn't get too far without delicious carbs, just about every plant is partially green. This isn't true across the board, though. Some plants use different pigments for photosynthesis, and there are a few hundred parasitic plant species that don't need chlorophyll because they mooch carbohydrates that are produced by other plants. But for the most part, land plants depend on chlorophyll to maintain their active plant lifestyle. And by extension, so do we, since animal life depends on plants to survive.

But why reflect green and absorb red and blue rather than the other way around? The short answer is that the red and the blue light are the good stuff. The sun emits more red photons than any other color, and blue photons carry more energy than other colors. Sunlight is abundant enough that it wouldn't be efficient to absorb all light, so plants evolved to absorb the areas of the light spectrum that offer the best bang for the buck. And it's a good thing, too: If plants needed to absorb the full spectrum of sunlight to get by, they would all be black. And the outdoors would have a decidedly gloomy tint.

Is Yellowstone National Park about to explode?

About three million people visit Yellowstone National Park each year. They do a little hiking, maybe some fishing. They admire the majesty of the mountains and antagonize a few bears for the sake of an interesting picture. And, of course, they visit the geysers. Hordes of tourists sit and wait patiently for Old Faithful to do its thing every ninety minutes or so. When it finally blows, they break into applause as if they've just seen Carol Channing belt out "Hello, Dolly." And then they go home.

Few of these tourists give much thought to what is going on below their feet while they are at Yellowstone. Geologists, however, have known for years that some sort of volcanic activity is responsible for the park's strange, volatile, steamy landscape. Just one problem: They couldn't find evidence of an actual volcano, the familiar cone-shaped mountain that tells us in no uncertain terms that a huge explosion once took place on that spot.

In the 1960s, NASA took pictures of Yellowstone from outer space. When geologists got their hands on these pictures, they understood why they couldn't spot the volcano: It was far too vast for them to see. The crater of the Yellowstone volcano includes practically the entire park, covering about 2.2 million acres. Obviously, we're not talking about your typical, garden-variety volcano. Yellowstone is what is known as a supervolcano.

There is no recorded history of any supervolcano eruptions, so we can only use normal volcanic activity as a measuring stick. Geologists believe that Yellowstone has erupted about 140 times in the past sixteen million years. The most recent blast was about one thousand times more powerful than the 1980 eruption of Mount St. Helens in Washington, and it spread ash over almost the entire area of the United States west of the Mississippi River. Some of the previous Yellowstone eruptions were many times more destructive than that.

And here's some interesting news: In the past twenty years or so, geologists have detected significant activity in the molten rock and boiling water below Yellowstone. In other words, the surface is shifting.

Nearby, the Teton Range has gotten a little shorter. Scientists have calculated that Yellowstone erupts about every six hundred thousand years. And get this: The last Yellowstone eruption took place about 640,000 years ago.

Before you go scrambling for the Atlantic Ocean, screaming and waving your arms in the air, know that the friendly folks who run Yellowstone National Park assure us that an eruption is not likely to happen for at least another thousand years. And even then, any eruption would be preceded by weeks, months, or perhaps even years of telltale volcanic weirdness.

So don't worry. It's safe to go to Yellowstone. For now. But go easy on the bears, okay? Photography may be your favorite hobby, but theirs is mauling.

What's the smelliest thing on earth?

Perhaps the skunk gets a bad rap. When someone wants to describe an object—or perhaps an acquaintance—as stinking up the place, the poor skunk is invariably used as the reference point.

It's true that the *Guinness Book of World Records* lists butyl seleno-mercaptan, an ingredient in the skunk's defense mechanism, among the worst-smelling chemicals in nature. But according to scientists and laboratory tests in various parts of the world, there are far fouler odors than a skunk's spray. Some of the most offensive nose-wrinklers are man-made. Dr. Anne Marie Helmenstine, writing in *Your Guide to Chemistry,* suggests that a couple of molecular compounds—which were invented specifically to be incredibly awful—could top the list.

One is named Who-Me? This sulfur-based chemical requires five ingredients to produce a stench comparable to that of a rotting carcass. Who-Me? was created during World War II so that French resistance fighters could humiliate German soldiers by making them stink to high heaven. The stuff proved almost as awful for its handlers, who found it difficult to apply so that they, too, didn't wind up smelling like dead flesh.

For commercial craziness, consider the second compound cited by Helmenstine. American chemists developed a combination of eight molecules in an effort to re-create the smell of human feces. Why? To test the effectiveness of commercially produced air fresheners and deodorizers. Ever imaginative, the chemists named their compound U.S. Government Standard Bathroom Malodor.

For many people, cheese comes to mind when thinking of man-made smells that make the eyes water. There is, in fact, an official smelliest cheese—a French delight called Vieux Boulogne. Constructed from cow's milk by Philippe Olivier, Vieux Boulogne was judged the world's smelliest cheese by nineteen members of a human olfactory panel, plus an electronic nose developed at Cranfield University in England. London's *Guardian* newspaper insisted that Vieux Boulogne gave off an aroma of "barnyard dung" from a distance of fifty meters.

A skunk would have a hard time matching that, and Pepe Le Pew might even take a backseat to the Bombardier beetle. This insect is armed with two chemicals, hydroquinone and hydrogen peroxide. When it feels threatened, the chemicals combine with an enzyme that heats the mixture. The creature then shoots a boiling, stinky liquid and gas from its rear. Humans unfortunate enough to have endured the experience claim that there's nothing worse.

No less a luminary than nineteenth-century naturalist Charles Darwin allegedly suffered both the smell and sting of the Bombardier beetle's spray when, during a beetle-collecting expedition, he put one in his mouth to free up a hand. Consider Darwin a genius if you like, but his common sense left something to be desired.

Why don't we run out of water?

Because Earth is one big water storage and recycling system. The amount of water on the planet is more or less constant—it's just continually changing form. The process is known as the water cycle.

The water cycle includes a variety of different paths, but it basically goes like this: When the sun heats the oceans, lakes, and rivers, water evaporates from their surfaces and forms water vapor. The process of evaporation eliminates the salt and impurities from seawater, leaving clean freshwater in gaseous form.

Some water vapor rises high in the atmosphere, cools, and condenses into tiny liquid droplets and ice that form clouds.

Sometimes the liquid droplets and ice in the clouds grow big enough to fall as rain, sleet, snow, and hail. The portion of this precipitation that collects on land soaks into the ground and flows into lakes, streams, and rivers, eventually making its way back to the oceans.

We use this water to sustain life, of course, and to keep our Slip 'n Slides slick and our cars shiny. But when we use water, we're usually just passing it through something (like our bodies) or adding stuff to it. We're not changing the actual water molecules, which remain part of the water cycle.

It's a pretty good system, but there's a catch: Yes, there is a massive, constant volume of water on the planet, but the vast majority is tied up in storage at any one time. Most of it exists in some form that is of little use to us—about 97 percent of Earth's water is undrinkable saltwater and about 2.1 percent is frozen in glaciers and icecaps. That leaves only about 0.9 percent in freshwater form, and much of that is underground, inaccessible to us. So while nature takes care of continually replenishing the freshwater supply, it leaves us a limited amount to use at any one time.

This is a problem because we're dangerously close to exceeding the rate at which nature can recycle water, even with help from modern water treatment facilities. In a sense, nature is up against a manmade consumption cycle: Modern agriculture and industry pollute a lot of water, which reduces the freshwater supply, while at the same time the demand for water is increasing as Earth's population continues to grow—and this growth compels agriculture and industry to expand.

Water shortages are at crisis levels in parts of Africa and Asia. Many scientists believe that the United States and much of the rest of the world will be in a similar predicament within fifty years unless we make some major changes. In this frightening scenario, the planet won't run out of water in the long-term—but it might in the short-term. If this comes to pass, a dry Slip 'n Slide will be the least of our problems.

Why does a red sky freak out sailors?

We learned the rhyme in elementary school:

> Red sky at morning, sailors take warning;
> Red sky at night, sailor's delight.

Unless your third-grade teacher was a meteorologist or a physicist, chances are you're still wondering what it means.

We'll tackle the physics first. Sunlight is made up of different colors—you've seen them all in a rainbow—and each color has its own wavelength. The blues are shorter and the reds are longer. The color that our eyes see depends on the path the light waves take and what happens to them along the way.

When the atmosphere is clean, air molecules do a good job of scattering the shorter light waves. Hence, we see a blue sky. When there's a bunch of dust and other particles—a.k.a.

aerosol—in the atmosphere, those particles scatter the longer light waves; then we see a red sky.

Now we'll do the meteorology part: Our weather systems are made up of alternating high- and low-pressure areas. If you're in a high-pressure system, a low-pressure system is on either side of you, and vice versa. High-pressure areas are usually dirty—lots of aerosols—and produce clear weather. Low-pressure areas are usually clean—far fewer aerosols—and produce inclement weather.

Here's what happens when the physics and meteorology are put together: As we look at the sky at sunrise or sunset, sunlight is traveling its longest path to our eyes during the course of a day, which means it's going though more atmosphere. If we are in a mid-latitude location, say somewhere in North America or Europe, most storms travel from west to east. If the atmosphere is dirty as the light makes this long journey to our eyes, the sky will look red when the light reaches our eyes. (Remember, high-pressure area = dirty atmosphere = scattered long wavelengths = red sky.)

So, when there's a red sky to the west, there's probably a high-pressure area to the west—clear weather's ahead. But if that red sky is in the east, the high-pressure area is also in the east and a low-pressure area is on its way—a storm is coming. (And, if we're far from mid-hemisphere, exactly the opposite applies.)

By the way, rainbows can send the same weather forecast, but for different reasons. Instead of aerosols in the atmosphere and scattered light, we see rainbows because of moisture in the atmosphere and refraction—or a change in the direction—of the light. The light from the sun refracts through the moisture in the clouds in the sky opposite the sun. A rainbow in the morning when the sun is in the east can signal rain moving eastward toward you. A rainbow in the evening when the sun is in the west means that the rain has already passed you.

So a red sky or a rainbow at sunrise tells sailors who are in the mid-hemisphere in the open water looking across a long distance to the horizon that a storm is likely moving toward them. That's why sailors grow concerned and call out, "Batten down the hatches!" At sunset, that red sky or rainbow can indicate there's no storm ahead. The sailors can relax and pass the grog.

Where does dew come from?

You wake up on a cool spring morning to find the sunrise glittering over the grass, refracted in a million tiny droplets of dew. To your dismay, your father's tool set, which you neglected to put away the night before, is also glittering in the sun. It didn't rain, but everything is wet. You run out, cursing the dew for possibly ruining his best ratchet set and hoping to get the tools put away before your dad wakes up. In the midst of your cursing, you wonder: If it didn't rain, where did all the water come from?

Dew gathers primarily on cool mornings, particularly during spring and fall, when the temperature is much lower than it was the previous evening. During the heat of the day, the air is filled with water vapor. As the air cools, the land (along with the grass, trees, flowers, and your dad's tool set) cools with it.

Water vapor becomes heavier as the temperature falls. As the weight increases, the air becomes oversaturated, and when that oversaturated air comes into contact with something cool—a blade of grass, a leaf, or a socket wrench (metal loses heat quickly)—water molecules cling together and form a dew droplet. Grass and plants are usually the first to collect dew because they lose water vapor themselves, making the air above them highly saturated with water. (Oversaturated air is also what gives us clouds, mist, fog, and rain. Small droplets of condensation form mist or fog, and larger droplets form rain.)

Dew will only gather on material that has cooled, however, which is why your driveway isn't wet in the morning—a concrete slab holds heat much longer than a blade of grass. The air above the driveway is warmed by the concrete, and thus not as heavily saturated as the air over bare ground.

Which is not to say that your dad's ratchet set will be safe if you leave it on the driveway instead of the lawn. Put that thing in the shed where it belongs.

Is it possible for the air temperature to change a hundred degrees in one day?

Yep, it's totally possible. As long as you're in Montana. Because while a hundred-degree rise or drop in temperature is extremely rare, it has happened at least twice since meteorologists started keeping records—and both times, it was in the Treasure State. When the weather turns on a dime, it's usually because of a collision of weather fronts—the boundaries between huge masses of air with different densities, temperatures, and humidity levels. And Montana happens to be ground zero in a perpetual weather-front war.

The biggest twenty-four-hour temperature swing on record occurred in Loma, Montana, on January 14-15, 1972. The thermometer went from -54 degrees Fahrenheit to 49 degrees, a change of 103 degrees. This barely beat the previous record, which had been set 190 miles away in Browning, Montana—on January 23, 1916, the temperature went from 44 degrees down to -56 degrees. Even though this is no longer the record for the biggest overall twenty-four-hour temperature swing, it's still the mark for the most dramatic *drop* in such a time period.

Montana owns the twelve-hour records, too. Temperatures in Fairfield, Montana, dropped 84 degrees between noon and midnight on December 14, 1924. And on January 11, 1980, the temperature in Great Falls, Montana, jumped 47 degrees in just seven minutes.

What makes the weather in Montana so volatile? It's all because of chinook winds—warm, dry air masses caused by high mountain ranges. Chinooks form when moist, warm air from the Pacific Ocean encounters the Rocky Mountains along Montana's western border. As an air mass climbs the western slopes of the mountain range, its moisture condenses rapidly, creating rain and snow.

This rapid condensation sets the stage for the chinook effect by warming the rising air mass. Then, as the air mass descends the other side of the mountain range, the higher air pressure at the lower altitude compresses it, making it even warmer. The result is an extreme warm front that can raise temperatures drastically in a short period of time. But the effect is often short-lived: Montana also is in the path of bitter Arctic air masses, so cold fronts sweep into the state just as warm air masses leave.

This raises the question: How the heck do Montanans decide what to wear when they get up in the morning?

Does anyone ever die in quicksand?

It's possible, but it hardly ever happens. In a battle between you and quicksand, you definitely have the advantage—even if you happen to be the black-hatted villain from a classic Western who totally deserves that slow, sandy death.

Quicksand is nothing more than ordinary sand that has been liquefied, usually by water that seeps up from underground. Why does a little water make such a difference? Normally, a sand dune can hold you on its surface because of the friction that the individual grains of sand exert on each other—when you step on the sand, the grains push on each other and, collectively, hold you up. But when the right amount of water seeps into a mass of sand, it lubricates each individual grain, greatly reducing the friction. If you agitate quicksand by walking through it, it acts like a thick liquid, and you sink.

This sandy sludge is denser than water, however—and much denser than your body. This means that you'll only sink waist-deep before you reach your natural buoyancy level; after that, you'll float. In other words, you won't gradually sink all the way to the bottom until only your hat remains, like that villain from the classic Western.

Still, it's fairly difficult to free yourself from quicksand. Once you're in, the quicksand settles into a thick muck around you. When you try to lift your foot, a partial vacuum forms underneath it, and the resulting suction exerts a strong downward pull. A 2005 study that was published in the journal *Nature* suggests that the force needed to pull a person's foot out of quicksand at one centimeter per second is equal to the force needed to lift a medium-size car. The best way to get free, according to this study, is to wriggle your arms and legs very slowly. This opens up space for water to flow down and loosen the sand around you, allowing you to gradually paddle to freedom.

Does all of this mean that you don't need to fear death by quicksand? Not exactly. There are still a number of ways to die in the sandy goo. First, if you really freak out, you could thrash around enough to swallow huge quantities of sand. Second, if you're carrying a heavy backpack or are inside of a heavy car, you could sink below the surface of the quicksand. Third, if you get stuck in quicksand near the ocean, you might not be able to free yourself before the tide comes in and drowns you. Finally, if you don't know the wriggling trick, you might just give up and expire of dehydration and boredom.

Are the colors of the rainbow always in the same order?

Yes. The order of the colors—red, orange, yellow, green, blue, indigo, and violet, from the top to the bottom—never changes. You may see a rainbow missing a color or two at its borders, but the visible colors always will be in the exact same order.

Rainbows are caused by the refraction of white light through a prism. In nature, water droplets in the air act as prisms. When light enters a prism, it is bent ever so slightly. The different wavelengths of light bend at different angles, so when white light hits a prism, it fans out. When the wavelengths are separated, the visible wavelengths appear as a rainbow.

The colors of a rainbow always appear in the same order because the wavelengths of the visible color spectrum always bend in the same way. They are ordered by the length of their waves. Red has the longest wavelength, about six hundred and fifty nanometers. Violet has the shortest, about four hundred nanometers. The other colors have wavelengths that fall between red and violet.

The human eye is incapable of seeing light that falls outside of these wavelengths. Light with a wavelength shorter than four hundred nanometers is invisible; we refer to it as ultraviolet light. Likewise, light with a wavelength longer than six hundred fifty nanometers cannot be seen; we call it infrared light.

Now, about that pot of gold at the end of a rainbow—how do you get to it? If we knew that, we'd have better things to do than answer these silly questions.

Do rivers always flow north to south?

No, rivers are not subject to any natural laws that compel them to flow north to south. Only one thing governs the direction of a river's flow: gravity.

Quite simply, every river travels from points of higher elevation to points of lower elevation. Most rivers originate in mountains, hills, or other highlands. From there, it's always a long and winding journey to sea level.

Many prominent rivers flow from north to south, which perhaps creates the misconception that *all* waterways do so. The Mississippi River and its tributaries flow in a southerly direction as they make their way to the Gulf of Mexico. The Colorado River runs south toward the Gulf of California, and the Rio Grande follows a mostly southerly path.

But there are many major rivers that do not flow north to south. The Amazon flows northeast, and both the Nile and the Rhine head north. The Congo River flaunts convention entirely by flowing almost due north, then cutting a wide corner and going south toward the Atlantic Ocean.

There's a tendency to think of north and south as up and down. This comes from the mapmaking convention of sketching the world with the North Pole at the top of the illustration and the South Pole at the bottom.

But rivers don't follow the conventions of mapmakers. They're downhill racers that will go anywhere gravity takes them.

How do we know that no two snowflakes are alike?

Well, do you know the Snowflake Man? In 1885, Wilson A. Bentley became the first person to photograph a single snow crystal. By cleverly adapting a microscope to a bellows camera, the nineteen-year-old perfected a process that allowed him to catch snowflakes on a black-painted wooden tray and then capture their images before they melted away.

A self-educated farmer from the rural town of Jericho, Vermont, Bentley would go on to attract worldwide attention for his pioneering work in the field of photomicrography. In 1920, the American Meteorological Society elected him as a fellow and later awarded him its very first research grant, a whopping twenty-five dollars.

Over forty-seven years, Bentley captured 5,381 pictographs of individual snowflakes. Near the end of his life, the Snowflake Man said that he had never seen two snowflakes that were alike: "Under the microscope, I found that snowflakes were miracles of beauty. Every crystal was a masterpiece of design and no one design was ever repeated."

Since Bentley's original observation, physicists, snowologists, crystallographers, and meteorologists have continued to photograph and study the different patterns of ice-crystal growth and snowflake formation (with more technologically advanced equipment, of course). But guess what? Bentley's snow story sticks.

Even today, scientists agree: It is extremely unlikely that two snowflakes can be exactly alike. It's so unlikely, in fact, that Kenneth G. Libbrecht, a professor of physics at Caltech, says, "Even if you looked at every one ever made, you would not find any exact duplicates."

How so? Says Libbrecht, "The number of possible ways of making a complex snowflake is staggeringly large." A snowflake may start out as a speck of dust, but as it falls through the clouds, it gathers up more than 180 billion water molecules. These water molecules freeze, evaporate, and arrange themselves into endlessly inventive patterns under the influence of endless environmental conditions.

And that's just it—snow crystals are so sensitive to the tiniest fluctuations in temperature and atmosphere that they're constantly changing in shape and structure as they gently fall to the ground. Molecule for molecule, it's virtually impossible for two snow crystals to have the exact same pattern of development and design.

"It is probably safe to say that the possible number of snow crystal shapes exceeds the estimated number of atoms in the known universe," says Jon Nelson, a cloud physicist who has studied snowflakes for fifteen years. Still, we can't be 100 percent sure that no two snowflakes are exactly alike—we're just going to have to take science's word for it. Each winter, trillions upon trillions of snow crystals drop from the sky. Are *you* going to check them all out?

Is there a way to make ocean water drinkable?

Water in the twenty-first century will be like oil was in the twentieth century: a precious, limited commodity that everybody needs and that is capable of driving nations to war. The difference is, the stakes will be even higher. You think living a week without gasoline would be tough? Try living a week without water. You'd be toast.

The United Nations estimates that by 2025, two-thirds of the world's population will not have access to adequate amounts of drinking water. Our "blue planet" is covered by vast oceans, but the freshwater we need for drinking makes up only a fraction of all the water. For various reasons, that supply of freshwater is dwindling.

In light of the looming water crisis, it seems ludicrous that we can't find a way to make those giant oceans drinkable. The good news is, we can; the not-so-good news is, it's expensive, and the impact on the environment is unknown.

Desalination (the process of removing salt from seawater) is not a new idea. Nearly two thousand years ago, crafty sailors distilled freshwater from seawater by boiling the seawater and collecting the runoff from the resulting vapor. (This is similar to the way nature creates fresh rainwater from ocean water.) But distillation on a large scale requires ridiculous amounts of energy.

A cheaper method of desalination is reverse osmosis, which involves pushing seawater through a membrane that filters out the salt. The energy cost for reverse osmosis is still about ten times higher than that of treating normal fresh-water, but encouraging developments that can make the process more cost-effective are ongoing.

One other issue threatens the viability of large-scale de-salination: How do we dispose of all of the leftover salt concentrate? A workable answer to that question has yet to emerge. Nevertheless, desalination is hardly a pie-in-the-sky solution to the planet's looming water woes. Approximately fifteen thousand desalination plants are in operation around the world, in places like California, Florida, the Middle East, and North Africa.

These plants provide significantly less than 1 percent of the planet's drinking water, so worldwide success is still a long way off. But we remain confident that the same civilization that proudly brought us the Clapper and the Hummer Limo can find a way to prevent us from dying of thirst.

Why does El Niño upset weather patterns?

El Niño, that problem child of climatology, has been blamed for disasters around the world: forest fires in southeast Asia, deadly floods in central Europe, tornadoes in Florida, mudslides in California, droughts in Zimbabwe, and devastating tropical storms in Central America.

What exactly is this atmospheric arch-villain, and where does it come from?

Named, ironically, for the Christ child, the scourge known as El Niño is not so much a single event as it is a predictably unpredictable combination of meteorological conditions that usually arrives in December. The disruptive patterns of El Niño appear roughly once every two to ten years, and to understand what El Niño does, you first have to consider what happens when it doesn't show up.

Ordinarily, during the closing months of the year, trade winds along the equator in the Pacific Ocean blow warm surface water westward, forming an immense warm pool northeast of Australia. At the same time, in the east, off the coast of Peru, cooler water rises to replace the warm water that has moved west. The warm pool in the west serves as a weather machine, pumping moisture into the atmosphere that generates storms all around the planet, in generally predictable patterns.

But some years, for reasons unknown to scientists, the trade winds never come. The warm pool never makes it to Australia, and the cool water never rises near Peru. Instead of occupying one spot, the weather machine spreads across a large span of the equatorial Pacific, and its unpredictable location means that the weather that it generates doesn't follow recognizable patterns. This creates a domino effect around the world, forcing a whole slew of atmospheric conditions to follow new, unusual patterns.

It doesn't end there. El Niño has an obstinate little sister, La Niña, that follows El Niño around and behaves just about as badly, but in direct opposition. As the effects of El Niño taper off, the trade winds pick up and blow even harder, pushing more warm water west than usual and pulling up an overabundance of cool water in the east. This turns all of the El Niño weather patterns inside out. Eventually, the trade winds stabilize and conditions around the world return to normal. The cycle is known to meteorologists as El Niño-Southern Oscillation, or ENSO.

It's tempting to blame specific weather events on El Niño, but the truth is that El Niño merely changes weather patterns—lots of other local conditions have to conspire to create an event like a mudslide or a hurricane. It's also tempting, in these days of heightened environmental awareness, to blame El Niño and the havoc it wreaks on global warming. But El Niños have been toying with the world's weather for at least 130,000 years, and while they've grown more frequent in recent times, that trend began some ten thousand years ago.

And as we all know from watching Fred Flintstone, driving a car in the Stone Age did not leave a carbon footprint—just a lot of harmless, three-toed footprints.

Freaky Food Facts

Why does Swiss cheese have holes?

Rumors continue to run rampant about this age-old question. Some say manufacturers allow mice to nibble on Swiss before packaging the cheese. Others insist crafty deli owners cut the holes by hand with their carving knives. However, both of these conspiracy theories have more holes than, well, Swiss cheese.

Truth be told, and it's a bit embarrassing, Swiss cheese has holes because it has bad gas. That's right. Those holes in your sweet, nutty Swiss are actually popped bubbles of carbon dioxide gas.

Where do these gassy bubbles come from? Well, all cheese begins with a combination of milk and starter bacteria. The type of bacteria used helps determine the flavor, aroma, and texture of the finished cheese product. In the case of Swiss, cheese-makers use a special strain of bacteria called *Propionibacter shermani*. During the curing process, when the cheese ripens, this *P. shermani* eats away at the lactic acid in the cheese curd, tooting carbon dioxide gas all the while.

Swiss cheese is a densely-packed variety with a thick, heavy rind, so this built-up gas has nowhere to go. Trapped inside, the gas forms into bubbles. These bubbles eventually pop, leaving behind the characteristic holey air pockets.

In formal cheese lingo, these holes are referred to as "eyes." And the art of cheese making is such that their sizes can be controlled. By adjusting acidity, temperature, and curing time, dairies can create a mild baby Lorraine Swiss with lacy-looking pinholes or a more assertive Emmentaler Swiss with eyes the size of walnuts.

Oddly, in the United States, the size of Swiss cheese holes is subject to United States Department of Agriculture regulation. Every wheel of Grade A Swiss that is sold in America must have holes with diameters between three-eighths and thirteen-sixteenths of an inch.

All of this goes to show that sometimes, it's best not to over-think your cheese. Just slap it on a cracker, pour a glass of wine, and enjoy.

Can you really fry an egg on the sidewalk?

Claiming that "it's so hot outside that you can fry an egg on the sidewalk!" is an exaggeration—unless you have the proper tools. Eggs must reach 144 to 158 degrees Fahrenheit to change from liquid to solid and be considered cooked, according to the American Egg Board. Even on the most searing summer days, the typical sidewalk falls way short of the 144 degrees necessary to get eggs sizzling and coagulating.

Pavement of any kind is a poor conductor of heat, says Robert Wolke in his book *What Einstein Told His Cook: Kitchen Science Explained.* For starters, when you crack an egg onto pavement, the egg slightly cools the pavement's surface. In order to fry an egg, the temperature of a sidewalk has to climb enough to start and maintain the coagulating process. Lacking a constant flame or source of heat from below or from the sides, pavement can't maintain a temperature that's hot enough to cook eggs evenly. Forget sunny side up—you're likely to end up with a runny mess.

But frying an egg on the sidewalk is not impossible. Just ask the contestants of the Solar Egg Frying Contest, held each Fourth of July in the little town of Oatman, Arizona. People come from far and wide in hope of winning a trophy for "the most edible" solar-cooked egg.

Technically, these people are cheating. Contestants in the Oatman egg fry are allowed the use of mirrors, magnifying glasses, aluminum reflectors, and any kind of homemade cooking surface or contraption they can devise to harness the power of the sun. It also seems that sidewalk cooking is a bit more plausible in Arizona in July: heat is high, humidity is low, and the liquid in the cooking eggs dries out a little faster.

Still, your best bet is to abandon a sidewalk altogether and use a surface similar to that of a frying pan. Spritz a little Pam onto the hood of your '57 Chevy and get cooking. Metal is a much better conductor of heat than concrete.

How do they salt peanuts in the shell?

No, bioengineers haven't created a super breed of naturally salty peanut plants (yet). The real answer isn't nearly as exciting.

To salt peanuts while they're still in the shell, food manufacturers soak them in brine (salty water). In one typical approach, the first step is to treat the peanuts with a wetting agent—a chemical compound that reduces surface tension in water, making it penetrate the shell more readily. Next, the peanuts are placed into an enclosed metal basket and immersed in an airtight pressure vessel that is filled with brine. The pressure vessel is then depressurized to drive air out of the peanut shells and suck in saltwater.

Peanuts may go through several rounds of pressurization and depressurization. Once the peanuts are suitably salty, they are rinsed with clean water and spun on a centrifuge in order to get rid of the bulk of the water. Finally, they are popped into an oven so that the drying process can be completed.

Now, if they could just figure out how to cram some chocolate in those peanuts.

Why do grapes spark in the microwave?

That's right—you can do more with your microwave than just pop corn and defrost chicken. You can cause a mini electrical storm, make a piece of fruit spontaneously ignite, and possibly create a floating plasma fireball. All you need is a bunch of grapes. Mind this caveat, however: These experiments can cause serious and even irreparable damage to your appliance. Do your family a favor—if you *must* see this for yourself, try the experiments in the microwave in the employee lounge at work, preferably while your boss is away on vacation.

Microwaves heat food by bombarding it with waves of mild radiation. When these waves come into contact with a substance that conducts electricity—such as the juice inside a grape—that substance becomes charged. If the grapes are placed close enough together, the charge will move back and forth between them. As the charge moves through the air, the air itself becomes charged, producing a light show known as "arcing."

In another variation on this experiment, a single grape is sliced in half, with a bit of skin left connecting the two halves. The electrical current produced in each section of grape will travel over this bridge, which will heat the skin and eventually cause it to catch fire. The point at which the grape starts to burn is known as "a good time to turn the microwave off." (See "serious and even irreparable damage," above.) If the reaction taking place is not stopped, the heated gas between the two grape halves could form a plasma fireball. This cloud of plasma is created by the microwave's electrical field, which feeds off the radiation inside the appliance. The radiation causes the cloud to get hotter and hotter. As this continues, the odds of your microwave surviving the experiment dwindle, so enjoy the show while you can.

Other foods have been known to produce the same effect, or a similar one: Cranberries, blueberries, and green peppers have all produced electrical arcs. Pickled cucumbers start to glow when they have been microwaved for a long enough span of time. But obviously, these experiments aren't recommended. Appliance manufacturers caution that running an empty microwave can cause major damage to the microwave tube. Running a microwave with a couple small pieces of fruit can cause the same kind of damage, if the machine is allowed to run long enough.

So wait until the boss leaves for Cancun, turn the lights down, and toss a couple grapes into the microwave. You may get fired, but at least Rod and Joey—that annoying duo from Accounting—will remember you as "the guy who got our microwave privileges taken away."

Why doesn't liquor freeze?

Remember the bottle of Jägermeister that was in your dad's freezer for close to fifteen years, always sloshing around, seemingly impervious to both temperature and time? Why didn't that bottle of booze ever freeze?

Actually, liquor does freeze, just at an extremely frigid temperature. According to research conducted at Purdue University, the freezing point of ethyl alcohol (the type found in alcoholic beverages) is –179.1 degrees Fahrenheit. Household freezers don't get anywhere near that cold, so they're not going to turn your liquor into a block of ice.

But what about the emergency six-pack of Budweiser that you keep stashed in your garage? Why does the beer turn to slush when the temperature drops into the twenties? Simple: A twelve-ounce can of Budweiser contains about 5 percent alcohol by volume. The other 95 percent or so of that tasty beverage is made up of water, as well as some hops, malts, and other flavorings. Water, of course, freezes at thirty-two degrees Fahrenheit. Because your Budweiser contains such a small amount of alcohol and such a large amount of easily freezable water, it will seem like the whole thing is frozen. As for the ancient bottle of Jägermeister in Dad's freezer? It contained enough alcohol–35 percent–to avert such blatant slushiness issues.

Think about that the next time you trot out to your garage on a winter evening and crack open a cold one.

Why does fruitcake keep so long?

That's easy: It's loaded with booze. No mold can grow in that much alcohol. Fruitcakes start with fruit—fresh or dried—that is typically soaked for a week in port wine, bourbon, or dark rum. When the cake batter is mixed, a cup of whisky, brandy, or another equally potent liquor is likely to be one of the ingredients.

Alcohol is sometimes added even after the cake is baked. Some recipes call for the fruitcake to be sprinkled with brandy once a week for a month or more. If you don't want to go the sprinkling route, you can soak a towel in brandy and wrap it around the cake. A few old-fashioned cooks bake the cake with a cup placed in the middle in order to create a deep depression. When the cake cools, the depression is filled with brandy or rum, which soaks into the cake. This process can be repeated again and again as the cake absorbs each dose of alcohol.

Not every cook wants to get the family drunk on fruitcake. Some recipes leave the booze out and substitute fruit juice. These teetotaler cakes can last a long time if they are stored in airtight tins, but alcohol is the key ingredient of a true fruitcake.

Early versions of fruitcake were carried on long campaigns by Roman legions and the Crusaders. Sometimes, fruitcake that was made from a past year's harvest was shared to invoke blessings on the current year's harvest.

Delayed consumption is a fruitcake tradition. A fruitcake that is given as a gift is rarely eaten right away. Indeed, "regifting" is a fruitcake ritual, and some of the most legendary fruit-cake creations last for decades.

How do they decaffeinate a coffee bean?

We've seen seedless watermelon, fat-free chocolate, and Emmy nominations for Charlie Sheen. Here's further proof that anything is possible: the naturally caffeine-free coffee bean.

Stop and read that again—not decaf coffee, but an all-natural, no-processing-necessary decaf coffee bean, straight from the plant. Scientists reported in 2004 that they had found three coffee plants growing in Ethiopia that contained almost no caffeine. And they hope that by the end of the decade, they will be able to develop a market-ready breed from caffeine-free coffee plants —meaning it won't be too long before you're pouring yourself a warm cup of natural decaf.

Until then, you'll have to drink coffee that's been decaffein-ated the old-fashioned way. Which raises the question: How do they remove caffeine from a coffee bean?

First, let's cover the basics. Decaffeinated coffee is at least 97 percent caffeine-free. When decaf was introduced in 1903, benzene was used in the caffeine-removal process. Yes, the same benzene that is a fuel additive and a known carcinogen. But fear not—this technique is a thing of the past. Contemporary caffeine-removal methods, called direct extraction and indirect extraction, are safer. Both processes begin with unroasted green coffee beans that are soaked in water or steamed so that the caffeine is soluble and ready for extraction. In essence, the caffeine is being worn down so that is can be more easily removed.

Direct extraction is aptly named because the beans come into direct contact with a decaffeinating agent after being softened. An example of this technique is the so-called European process, in which the softened beans are continually rinsed with methylene chloride, a solvent. (Methylene chloride is strictly regulated in both Europe and the United States because it is a carcinogen.) This causes the caffeine to leech out of the beans and form a solution with the methylene chloride. The beans are then removed from the solvent and given another steam to evaporate any remaining chemicals. Finally, the beans get torched: They're roasted at 450 degrees Fahrenheit, which burns off any solvent residue.

Indirect extraction, on the other hand, occurs when green coffee beans are soaked in a water-based solution called "flavor-charged water." This special water draws out the caffeine as well as a few other oils that are found in coffee. The water is treated with a decaffeinating agent and then is used to impart some flavor back into the decaffeinated beans. After they've been roasted, the beans are good to go.

These methods will recede into history once the naturally caffeine-free coffee bean takes over the marketplace. Still, the question will linger: Why would anyone drink decaffeinated coffee in the first place?

If beans cause gas, why can't we use them to power our cars?

There are two ways to consider this question. Taking the high road, we can discuss the technology that transforms biomass into ethanol, a proven fuel for cars. Beans, like corn or virtually any other organic material, contain starches and complex carbohydrates that can be refined into ethanol, a combustible alcohol blended with gasoline to become that "green" E85 fuel you've heard about. Through fermentation—and with a lot of help from science—beans and their organic cousins can also find their way into methane gas, another proven biomass automotive fuel.

But how boring is the high road?

By "causing gas," this question really refers to the process by which the consumption of beans produces that bloated feeling that escapes us as ... well ... flatulence. High-fiber foods tend to cause intestinal gas, but beans seem to bear most of the blame, maybe because other world-class gas-promoters like cabbage and Brussels sprouts aren't as big a part of our diet.

The culprit in these foods is a natural family of hard-to-digest sugars called oligosaccharides. These molecules boogie their way through our small intestine largely unmolested. The merrymaking begins when they hit the large intestine. Bacteria living there strap on the feedbag, chomping away at this nutritional bounty, multiplying even. Our intestinal gas is the by-product of their digestive action.

Most of this gas is composed of odorless hydrogen, nitrogen, and carbon dioxide. In some humans—about 30 percent of the adult population—this process also produces methane.

Ethanol isn't a part of the oligosaccharide equation. But hydrogen and methane are, and they're flammable gases. In fact, hydrogen is another player in the fuel-of-the-future derby and already powers experimental fuel-cell vehicles.

So order that chili and fill 'er up. Beans in your Beemer! Legumes for your Lexus!

No so fast, burrito boy. Setting aside the daunting biotechnical hurdle of actually capturing bean-bred flatulence from a person's, um, backside, the challenge becomes one of volume and storage.

Human flatulence simply doesn't contain hydrogen or methane in quantities sufficient to fuel anything more than a blue flame at a fraternity party. Even if we did generate enough of these gases to power a car, they'd have to be collected and carted around in high-pressure tanks to be effective as fuels.

Human biochemistry is a wonderful thing, but it isn't yet a backbone of the renewable-energy industry. For that, breathe a sigh of relief.

Is it blood that gives red meat its color?

You might think so, but no. The red color is the result of myoglobin, a richly pigmented protein that is fixed within the tissue cells. Myoglobin receives oxygen from the blood and transfers that oxygen to the animal's working muscles for energy. Muscles that are used frequently require more oxygen, so they contain more myoglobin. The more myoglobin there is in the cells, the redder or darker the meat is.

Even white meats like chicken and turkey contain myoglobin. However, the concentration of myoglobin pigment is not as heavy in poultry as it is in beef. Red-meat animals such as cattle need constant energy for standing, walking, and extended periods of activity, so they have higher levels of myoglobin in their muscles. The result is meat that has a much more intense coloration.

But here's the catch: Myoglobin is deep purple in color, not red. Immediately after being cut, meat is quite dark; it turns a bright cherry-red color after its surface comes in contact with oxygen. This reaction creates a pigment called oxymyoglobin, and that's the color most consumers associate with freshness.

Red meats at the grocery store are often packaged in plastic wrap that allows oxygen to pass through to the meat in order to maintain this pleasing crimson color. However, that brilliant oxymyoglobin pigment is highly unstable and usually short-lived. Grocery-store lighting and continued exposure to oxygen lead to the production of metmyoglobin—a pigment that turns red meat a much duller shade of brown that isn't the least bit appetizing.

Is chocolate an aphrodisiac?

This question has been debated for ages. Giacomo Casanova—the poster boy for carnal exploration—certainly thought that chocolate did the trick. The eighteenth-century Italian would drink the sweet stuff before embarking on his amorous adventures.

Modern-day scientists are less convinced that chocolate is an aphrodisiac, although many do concede that the confection contains some curious properties. They're found in chocolate's key ingredient, the cacao bean, and include:

· Phenylethylamine (PEA). PEA causes blood pressure and blood-sugar levels to rise, temporarily resulting in heightened alertness and a state of happiness.

· Caffeine. This strong, fast-acting stimulant increases the heart rate and gets the blood flowing to all the right places.

· Anandamide. A type of cannabinoid—yes, cannabinoids are also found in marijuana—anandamide produces feelings of contentment.

· Theobromine. This mild, long-lasting stimulant has a mood-lifting effect. It opens blood vessels and stimulates heart muscle tissue.

· Tryptophan. Tryptophan triggers a discharge of serotonin, which provides an overall sense of happiness.

In addition, the consumption of chocolate prompts the body to release endorphins. What's so great about endorphins? Plenty: They are natural opiates that lower a person's sensitivity to pain and deliver feelings of contentment.

So, yes, chocolate can make you happy, happy, happy. But does it enhance sexual desire? That's the million-dollar question—and no one's come up with a definitive answer. John Renner, the founder of the Consumer Health Information Research Institute, has an intriguing take on the matter: "The mind is the most potent aphrodisiac there is. It's very difficult to evaluate something someone is taking because if you tell them it's an aphrodisiac, the hope of a certain response might actually lead to an additional sexual reaction."

Why do Wint-O-Green Life Savers spark when bitten in the dark?

Looking to create some sparks on your next date? Pop a Wint-O-Green Life Saver into your mouth before going in for a kiss. With one bite, you'll have impeccably fresh breath–and your own mini light show.

Ah, the fireworks of young love! And isn't it romantic that what's going on here is quite literally the science of attraction? The blue-white sparks of Wint-O-Green Life Savers are created through a process called triboluminescence, or light generated though friction.

Here's how it works: When the Life Saver is bitten, the sugar crystals in the candy break into fragments that are positively and negatively charged. These charges tend to retreat to opposite sides, but just as they're being pushed away from each other, they decide that they want to get back together. To do this, they jump across the air and back into each other's arms–making, in essence, tiny lightning bolts.

Wint-O-Green Life Savers aren't the only candies that create sparks of triboluminescence. In fact, most hard, sugary candies–and even plain old sugar cubes–produce a glow of ultraviolet light when cracked. But most of the time, the light is too faint to be seen. Why do Wint-O-Green Life Savers produce a greater amount of visible light? The oil of wintergreen flavoring in the candy (methyl salicylate) is naturally fluorescent. The fluorescent oil converts nearly invisible ultraviolet light into a visible bright blue light, which adds to the triboluminescent effect.

To try this at home, find a dark room and bring a mirror. If you happen to be in the company of a date, here's a last bit of advice: Make sure this is the only time you chew with your mouth open.

Who created time zones?

On a pleasant July evening in 1876, Sir Sanford Fleming was waiting in a railroad station in Bandoran, Ireland, for a train that had been listed in his *Railway Travelers Guide* as due at 5:35. When the train failed to arrive, he inquired at the ticket office and learned that it stopped there at 5:35 in the morning, not 5:35 in the evening. Fleming might have just fired off an irritated letter to the editor of the *Guide;* instead, he decided it was time to change time.

Up to that point in history, the sun had ruled time. Earth rotates at approximately 17.36 miles per minute, which means that if you move thirty-five miles west of your present location, noon will arrive about two minutes earlier. Going the same distance east, it would come two minutes later. Confusing? Yes. But back in horse-and-buggy days, keeping precise track of time wasn't really an issue. What difference did a few minutes make when your only goal was to arrive at your destination before sundown?

The invention of the railroad altered this ancient perception of time forever. To run efficiently, railroads needed a schedule, and a schedule needed a timetable, and every minute did indeed count. Fleming, who had worked as a railroad surveyor in Canada, was even more aware of the confusion over time than most people. Each railroad company used its own time, which was set according to noon at company headquarters. A weary traveler might be faced with five or six clocks at the station. Which one was correct?

Fleming came up with what he believed to be an ingenious solution. Earth would be divided into twenty-four sectors, like the sections of an orange, each fifteen degrees latitude apart. Each section would become a time zone, its clocks set exactly one hour earlier than the preceding zone.

Though Fleming's proposal was a model of common sense, he had a hard time convincing people to buy into it. The United States was an early adaptor, mandating four continental time zones in 1883. A year later, President Chester Arthur assembled the International Prime Meridian Conference in Washington, D.C. Twenty-five nations were invited and nineteen showed. They chose the Royal Observatory at Greenwich, England, as the prime meridian because it was already used by the British Navy to set time.

It wasn't until 1929, however, that standard time zones were instituted throughout the world. Fleming also proposed the use of a twenty-four-hour clock, which would have meant that his evening train would have been scheduled to arrive at 17:35 rather than 5:35. This never caught on, except in the military and hospitals.

The sun remains our touchstone when it comes to time. We still recognize the twin poles of noon and midnight—one light, the other dark. Each, however, has the same number affixed to its name, which reminds us that on this planet, what goes around will always come around again.

Who invented the match?

For thousands of years, "keep the home fires burning" wasn't a cute saying—it was a major undertaking. Once your fire went out, there was no way to start it again except with good old-fashioned friction (i.e., rubbing two sticks together or striking a flint against a rock until you got a light).

Around 1680, Robert Boyle, an chemist from Ireland, discovered that a stick coated with sulfur would ignite instantly when rubbed against a piece of paper coated with phosphorous. But prior to the Industrial Revolution, both sulfur and phosphorous were expensive and hard to produce, so Boyle's discovery had no practical application for nearly one hundred and fifty years.

Real matches appeared on the market in 1827 after John Walker, an English chemist and apothecary, stirred up a mixture of potassium chlorate and antimony sulfide. He coated the end of a stick with this mixture, let it dry, scraped it against sandpaper, and—just like that—fire. Walker named his matchsticks Congreves, after the weaponry rockets that were developed by Sir William Congreve in about 1804. Like rockets, Walker's Congreves often did more harm than good, sending out showers of sparks that lit not just lamps and stoves, but rugs, ladies' dresses, and gentlemen's wigs.

But such calamities didn't deter Samuel Jones, another Englishman. Jones modified Walker's process to make it less explosive, patented the result, and called his product Lucifers, a playful reference to the devilish odor given off by burning sulfide. Despite their nasty stench, Lucifers proved to be a big hit among gentlemen who liked to indulge in the new pastime of smoking cigars.

In an effort to produce an odor-free match, French chemist Charles Sauria added white phosphorous to the sulfur mixture in 1830. Unfortunately, white phosphorous not only killed the smell, but it also killed those who made the matches. Thousands of the young women and children who worked in match factories began to suffer from phossy jaw, a painful and fatal bone disease caused by chronic exposure to the fumes of white phosphorous. (Once white phosphorous was understood to be poisonous, reformers worked to ban it from matches, finally succeeding with the Berne Convention of 1906, an international treaty that prohibited its use in manufacture and trade.)

In the 1850s, Swedish brothers John "Johan" and Carl Lundstrom created a match that was coated with red instead of white phosphorous on the striking surface. Red phosphorous was more expensive, but unlike its pale cousin, it was nontoxic when inhaled.

Over the next sixty years, inventors experimented with many types of red phosphorous matches, the best being the "safety" matches that were patented by the Diamond Match Company of the United States in 1910. President William Howard Taft was so impressed by the company's new matches that he asked Diamond to make its patent available to everyone "for the good of all mankind." On January 28, 1911, Diamond complied, and ever since, the match business has been booming. You might say it's spread like wildfire.

Why is IBM referred to as Big Blue?

From its inception in 1911 as the Computing-Tabulating-Recording Company, International Business Machines Corp. has been perhaps the most powerful and sophisticated force behind society's obsession with the collection and storage of data. It would seem simple, then, to find out how IBM got its nickname, Big Blue. Surely somewhere inside this behemoth of a corporation, someone must have recorded the origins of the company's nickname and stored it in some special database for all eternity. Nope.

Even though IBM sometimes refers to itself as Big Blue and has incorporated "Blue" into the names of some of its products (Deep Blue, Blue Pacific, Blue Gene), the Armonk,

New York-based firm can't definitively explain the source of its nickname. Of course, there are theories. Some sleuths believe that people started calling IBM Big Blue because the company's employees were required for many years to wear white shirts, which prompted a number of them to wear blue suits.

Perhaps a more cogent explanation is that the IBM logo has incorporated the color blue since the 1940s. The most plausible theory might be that the mainframes IBM sold in the 1960s had blue covers, which led sales reps and customers to coin the term Big Blue. Business writers picked up the term and popularized it.

As we move into our second century of high-tech data management, Microsoft has assumed the mantle that IBM once held as the king of business technology. Maybe someday it'll get a nickname. How about Not-So-Big Blue?

How have the Russians preserved Lenin's body?

Lenin lived, Lenin lives, Lenin will live.

–Soviet slogan

On January 21, 1924, Soviet leader Vladimir Lenin died after a series of strokes. Two days later, pathologist Alexei Abrikosov embalmed Lenin's body so it would be presentable for viewing. A makeshift wooden tomb was designed and built by architect Alexey Schusev on Red Square by the Kremlin Wall in Moscow. More than one hundred thousand people visited the tomb within a month and a half.

By August, people were still coming to pay their respects to the man who had spearheaded the Bolshevik Revolution in Russia and brought about Communist rule. Scientists Vladimir Vorbiov, Boris Zbarsky, and others routinely reintroduced the preservative chemicals into Lenin's body to keep it from putrefying.

Joseph Stalin, Lenin's successor, perhaps sensing that the fervor attached to Lenin could be harnessed for his own purposes, created the Commission for Immortalization, and the decision was made to preserve Lenin's body until the end of the Soviet state—presumably forever.

Lenin is kept in a glass coffin at Lenin's Mausoleum on Red Square. The only visible parts of his body are his head and his hands. (Stalin was placed in the same tomb upon his death in 1953, but was removed in 1961 by then-Soviet leader Nikita Khrushchev.) Lenin wears a plain suit, and the lower half of his body is covered with a blanket.

Vorbiov and Zbarsky devised a permanent embalming technique: Every eighteen months, Lenin's body is immersed in a glass tub of a solution of glycerol and potassium acetate. The chemicals penetrate his body, and he becomes like any living human insofar as 70 percent of his body is liquid. After he's taken out of the tub, Lenin is wrapped in rubber bandages to prevent leakage. He is then dressed and groomed. Occasionally, a bacterial growth will develop, but it is quickly scrubbed off.

Lenin has remained in his tomb since 1924, except for a brief evacuation to Siberia during World War II when it looked as if the Nazis might take over Moscow. Lenin's tomb is one of Moscow's main tourist attractions. The Soviet state is long gone, having collapsed in 1991, but its father lives on. Sort of.

What are the ingredients of Love Potion No. 9?

We thought we should ask, just in case you're looking to hook up with that cute barista at the corner coffee shop. Love potions have long been credited with having major magical influences over the whims and woes of human attraction. And they just might work.

In the second century AD, Roman writer and philosopher Apuleius allegedly concocted a potion that snagged him a rather wealthy widow. Relatives of the widow even brought Apuleius to court, claiming the witchy potion had worked to subvert the woman's true wishes. Apuleius argued that the potion (supposedly made with shellfish, lobsters, spiced oysters, and cuttlefish) had restored his wife's vivacity and spirit—and the court ended up ruling in his favor.

Yes, love potions have been the stuff of history and mystical legend since ancient times. These alluring elixirs played a major role in Greek and Egyptian mythology, and even made an appearance in the 2004 fairy-tale flick *Shrek 2*. In the movie, the Fairy Godmother gives the King of Far Far Away a bottled potion intended to make Fiona fall in love with the first man she kisses.

That bottle was marked with a Roman numeral IX, by the way, a clear nod to the formula first made famous in the doo-wop ditty "Love Potion No. 9," which was recorded by The Clovers in 1959 and The Searchers in 1963. According to the song, as penned by legendary songwriters Jerry Leiber and Mike Stoller, the ingredients for the concoction "smelled like turpentine, and looked like Indian ink."

Doesn't sound too appealing, huh? Well, it apparently did enough to help a guy who was "a flop with chicks." That is, until he "kissed a cop down on 34th and Vine."

At any rate, if you're a forlorn lover looking to make a little magic of your own, you just might be in luck. In the mid-1990s, Leiber and Stoller worked with former guitarist and part-time perfumer Mara Fox to develop a trademarked cologne spray bearing the name of their hit song.

According the label, Love Potion #9 is made with water, SD40B alcohol, isopropyl myristate, isopropyl alcohol, and the fragrances of citrus and musk. Can this cool, clean scent really heighten your passion and arousal and make you attractive to the opposite sex?

Hey, if George Clooney or Angelina Jolie happens to be in the area, it's certainly worth a spritz! However, the perfume does come with a disclaimer: "No guarantee of success is granted or implied."

Who invented the brown paper bag?

Consider the familiar, flat-bottomed brown paper bag: It's useful, ubiquitous, and utterly simple. Now get a sheet of brown paper, a pair of scissors, and some glue, and try to make one yourself. Not so simple, huh?

In 1870, Margaret Knight of the Columbia Paper Bag Company in Springfield, Massachusetts, was doing the same kind of puzzling over paper bags. Back then, the only paper bags that were being manufactured by machine were the narrow, envelope kind, with a single seam at the bottom. Flimsy and easily broken, they were despised by merchants and shoppers alike. The paper bag business was not booming. So Maggie Knight set out to build a better bag.

Born in 1838, Knight had been tinkering with tools since childhood; while other girls played with dolls, she excelled at making sleds and kites. She was especially fascinated by heavy machinery. At the age of twelve, she invented a stop-motion safety device for automatic looms after witnessing an accident in a textile mill that nearly cost a worker his finger. Though never patented, her invention was widely employed throughout the industry.

During her twenties and early thirties, Knight tried her hand at several occupations before finally landing at Columbia Paper Bag. Working alone at night in her boarding house, she designed a machine that could cut, fold, and glue sheets of paper into sturdy, flat-bottomed bags. This time, she applied for a patent. On July 11, 1871, Patent No. 116842 was issued to Margaret E. Knight for a "Bag Machine." Her employer was eager to implement her design, but the male workers he hired to build and install the new machines refused to take direction from a woman, until they were convinced that Maggie was indeed the "mother" of this particular invention.

Knight also had to fend off a challenge to her patent by a rival mechanist, who had spied on the construction of her first proto-type. The court decided in her favor, and she persisted in her career. After leaving Columbia, she cofounded the Eastern Paper Bag Company in Hartford, Connecticut, and also supervised her own machine shop in Boston. Between 1871 and 1911, she received twenty-six patents in her own name and is thought to have contributed to more than fifty inventions patented by others; she also built scores of unpatented devices. Upon her death in 1914, the press lauded her as America's "female Edison."

Among her most successful inventions were an easy-to-install window frame, a number-stamping machine, and a mechanical roasting spit. The humble paper bag, however, remains her greatest contribution to civilization. Even today, bag manufacturers rely on her basic concept. So the next time you decide to brown-bag your lunch, stop and give thanks to Maggie Knight for the paper bag—useful, ubiquitous, and a work of genius, too!

When will jet packs be ready for consumer use?

Jet packs no longer are just the stuff of James Bond movies—they are a reality. But you need to be rich—part of the jet set, so to speak—to get one.

The jet pack, also known as a rocket belt, has a simplistic aim: It enables users to avoid clogged streets by taking to the sky. It takes a sophisticated design to accomplish this feat. The device is strapped to the wearer's back and propels him or her off the ground using small rockets.

Much of the jet pack's development is credited to Bell Aerosystems engineer Wendell Moore, who designed the Small Rocket Lift Device in 1953. His shaky, unstable prototype ran on nitrogen gas and generated enough interest to secure a contract from the United States Army. Moore's design evolved to include a 280-pound thrust rocket motor that ran on peroxide, but it only enabled users to stay aloft for only twenty-one seconds. As a result, the Army eventually lost interest. Many subsequent versions were designed, some with wings, but each struggled to address the same challenges, such as flight time, user safety, fuel, and weight.

In 2007, two companies introduced consumer-ready jet packs with lift times approaching nineteen minutes. Colorado-based JetPack International uses standard jet fuel in its model, and Tecnologia Aeroespacial Mexicana sells a rocket belt that runs on propane. JetPack's model, the T 73, goes for about $200,000, including training.

Both manufacturers have had difficulty finding buyers for their products, which are not user-friendly in addition to carrying hefty price tags. Indeed, jet packs are so trivial that the Federal Aviation Administration has yet to institute licensing regulations for the contraptions.

Who invented the computer mouse?

Douglas Engelbart. And here's an extra-credit question: When did he do it? Is your guess the 1990s or the 1980s? If so, you're wrong. The correct answer is the 1960s.

Engelbart grew up on a farm, served in the Navy during World War II, and then obtained a PhD. He wound up at the Stanford Research Institute, where he pursued his dream of finding new ways to use computers. Back in the 1950s, computers were room-filling behemoths that fed on punch cards.

Engelbart believed that computers could potentially interact with people and enhance their skills and knowledge; he imagined computer users darting around an ethereal space that was filled with information. Most folks couldn't envision what he had in mind—no one thought of a computer as a personal machine, partially because the models of the day didn't even have keyboards or monitors.

Engelbart set up the Augmentation Research Center lab in 1963 and developed something he called the oNLine System (NLS). Today, we would recognize NLS as word-processing documents with hypertext links, accessed via a graphical user interface and a mouse.

On December 9, 1968, after years of tinkering, Engelbart presented his new technology at the Fall Joint Computer Conference in San Francisco. Engelbart's mouse—with a ball, rollers, and three buttons at the top—was only slightly larger than today's models. It was called a mouse because of its tail (the cord that connected it to the computer), though no one remembers who gave it its name. We do know, however, that engineer Bill English built the first mouse for Engelbart.

To Engelbart's disappointment, his new gadgets—including the mouse—didn't immediately catch on. Some curmud-geons in the audience thought that the ninety-minute demo was a hoax, though it received a standing ovation from most of the computer professionals in attendance. Eventually, Engelbart got the last laugh: His lab hooked up with one at UCLA to launch the ARPANET in October 1969. ARPANET, as any computer geek knows, was a precursor to the Internet.

Engelbart's mouse patent expired in 1987, about the time the device was becoming a standard feature on personal computers. Consequently, he has never received a dime in royalties. But he was never in it for money—Engelbart's motivation was to raise humanity's "Collective IQ," our shared intelligence.

In November 2000, President Clinton presented Engelbart with the National Medal of Technology, the highest honor the nation can award a citizen for technological achievement. Engelbart may not have gotten rich from his computer mouse, but he at least gained a measure of lasting fame.

Did Ben Franklin really discover electricity?

If by "discover," we mean to ask if Ben Franklin was the first to stumble upon electricity, the answer is no. But there's more to the story.

The ancient Greeks knew about static electricity. They observed amber, which they called *elektron,* drawing feathers and twigs to itself when rubbed with a cloth. Static electricity remained an amusing trick until the eighteenth century, when an Englishman named Stephen Gray found that electricity could be "sent" hundreds of feet if the right material was used to conduct it. Dutch professor Pieter van Musschenbroek of Leyden figured out that electricity could be stored in water-filled jars. He realized this after he touched a wire that was sticking out of a jar and accidentally electrocuted himself. He survived, and Leyden jars became the rage among scientists.

News of Leyden jars and electrical displays eventually reached the intellectual backwater of North America, where wealthy printer Ben Franklin saw a demonstration and became fascinated. He ordered jars and scientific journals from London and set up his own experiments.

Here's where it gets a bit tricky. There's a second definition of "discover": to obtain knowledge of something through study. In this sense, Franklin did indeed discover a lot about electricity. Tricks aside, he was determined to find a real, practical use for the strange energy. In pursuit of this audacious goal, Franklin was knocked senseless while arranging to use electricity to kill and roast a turkey.

Like other scientists, Franklin suspected that lightning was pure electricity. Two years after his accident, Franklin and his son launched a specially built silk kite into a stormy sky. The kite was not, in fact, struck by lightning. By simply remaining airborne under storm clouds the kite was electrified—and so was the metal key that was tied to its tail near Franklin's hand.

It seems simple now, but figuring out a way to prove that lightning was electricity made Franklin famous on both sides of the Atlantic. Franklin, while still in his forties, received honorary degrees from Harvard and Yale and a gold medal from London's Royal Society. He was a celebrity—and the American Revolution hadn't even started.

How do those transparent TelePrompTers work?

For those of you too busy watching Cartoon Network to be bothered with distractions like the State of the Union address, the technology in question is the seemingly transparent panes of glass flanking the president during a speech. Though this style of TelePrompTer has been used in other events, it has become so synonymous with speeches given by the president that it is colloquially known as a "presidential TelePrompTer." What kind of sorcery can deliver the content of a speech to a speaker without being seen by the audience?

While these TelePrompTers look like they're from the future, they are actually a modification of the type of system that is used in just about every TV newscast. With normal TelePrompTers, a monitor is placed perpendicular to the lens of a camera that displays the text the anchor reads. Then a half-silvered mirror—a piece of glass or plastic with a very thin reflective layer on one side, similar to the one-way mirror you'd find in an interrogation room—is placed at an angle in front of the camera lens and above the monitor. The mirror reflects the words on the monitor to the speaker but is transparent to the camera. The result is the illusion that the newscaster is staring directly into the camera, delivering line after line of flawless copy, apparently from memory.

The presidential TelePrompTer is a slight modification of this method. For this setup, two flat-screen LCD monitors are placed on the podium; they reflect the text upward onto two treated squares of glass that are like the aforementioned half-silvered mirror.

The next time you see a speech in which a presidential Tele-PrompTer is used, notice how the glass is tilted toward the podium to catch the monitor's reflections instead of away from the speaker—it's the opposite of how a music stand, for instance, would be positioned to hold a sheet of paper. The result is that the glass panels appear blank to the audience while the speaker sees the reflected text from the LCD monitors. The glass is placed on both sides of the president so that our fearless leader won't miss a beat while "naturally" turning left and right to address both sides of the room.

So it's not magic at all, just boring old science. Sorry. Now you know why magicians never reveal the secrets behind their tricks.

What makes superglue stickier than regular glue?

The short answer is that superglue forms an actual chemical bond between molecules, while the stickiness in normal glue is caused by much weaker attractions between molecules.

Any glue—from the stuff you used (and sometimes ate) in kindergarten to the space-age superglue that's stuck to the bottom of your junk drawer—has two types of stickiness. It needs to be adhesive, meaning that the molecules stick to other material. It also needs to be cohesive, meaning that the molecules stick to each other.

In primitive kindergarten glue, the adhesion and cohesion depend on a set of phenomena known as the van der Waals forces—weak, attractive, or repulsive forces that arise between molecules. (That vowel-heavy name is courtesy of Johannes Diderik van der Waals, the Dutch scientist who proposed the existence of intermolecular forces in the 1870s.) These forces include the attraction between adjoining molecules caused by slight differences in electrical charges on either side of each molecule.

The van der Waals forces can make the molecules of regular glue adhere and cohere, but only weakly. There's no chemical reaction to interlace the molecules and create a truly durable bond. But if there's no chemical reaction, what's going on when regular glue dries and hardens? Good question.

Regular glue is a liquid because it's an emulsion of a sticky compound in water. When glue is in a bottle, the water molecules keep the compound molecules flowing so that they can't stick firmly together. But when you squirt the glue out of the bottle and expose it to air, the water evaporates. The remaining molecules of the sticky compound cling together and to any other dry material that the glue touches, making things stick to each other.

Superglue is based on an entirely different principle: a chemical reaction that forms chemical bonds. The main ingredient in superglue is cyanoacrylate, a chemical compound that polymerizes when exposed to water. In other words, the hydrogen and oxygen atoms in water enable the cyanoacrylate molecules to form chemical bonds with one another. These molecules bind together and to whatever they're touching, forming a hard plastic. These chemical bonds are much stronger than the van der Waals attractions between molecules in normal glue.

The little bit of moisture on most surfaces is enough to trigger this reaction, which is why superglue sets so quickly on just about anything. This is also why it's much more dangerous to eat superglue than regular glue: It will form a hard plastic in your body. So if you're so in touch with your inner child that you can't resist tasting glue now and then, you should stick to standard paste.

How smart do you have to be to be considered a genius?

We all know a moron when we see one, but geniuses can be a bit harder to pick out of a lineup.

The simplest gauge of mental capacity is intelligence quotient, or IQ. Different IQ tests use different types of questions, but they all share a basic scoring system called the Binet Scale (after French psychologist Alfred Binet, who came up with it in the early nineteen hundreds).

In the modern version of the Binet Scale, 100 is the median score, and "average intelligence" is any score between 90 and 109. Scores of 110 or higher indicate superior intelligence, and scores of 140 or higher mean truly exceptional intelligence—less than 1 percent of the population is in this rarefied category.

In the early twentieth century, researchers who studied intelligence in children began setting a genius benchmark, typically between 130 and 140. They assumed that with this kind of brainpower, these child geniuses would be smart enough to succeed at just about any mental task. The definition caught on with parents and psychologists, and 140 became the most common magic number. So if you crack this barrier, you have a defendable rationale for sporting that "Genius at Work" T-shirt.

But let's not get ahead of ourselves—today, many psychologists believe that IQ in itself is an incomplete, and perhaps even a flawed, measure of intelligence. This is partly because of a concept that was first advanced in 1983, when psychologist Howard Gardner rocked the profession with his theory of multiple intelligences. Gardner posited that people exhibit seven separate types of intelligence: linguistic, logical-mathematical, spatial, musical, bodily-kinesthetic, interpersonal, and intrapersonal. Conventional IQ tests are focused on linguistic and logical-mathematical intelligence. They are, then, a bad measure of other types of intelligence; for example, an IQ test won't reveal artistic or athletic genius.

Another problem with IQ tests, at least where we geniuses are concerned, is that the questions don't measure innovation—the ability to make completely original connections, or what might be called "flashes of genius." Many people with very high IQs end up doing nothing remarkable their whole lives. Does it make sense to call them geniuses, just because they can answer test questions that have established, known answers? As smart as Einstein was, he never would have gone down in history as a genius if he hadn't come up with new ideas.

A more useful, but much more subjective, definition of "genius" is a person with exceptional intelligence (of any sort) that enables him or her to make creative leaps that change how we see or interact with the world. It might be a new way of painting (Picasso's cubism), a new way of explaining natural phenomena (Newton's laws of motion), or an amazing invention (Ron Popeil's Veg-O-Matic).

So, Mr. Smarty Pants, even if you do have a sky-high IQ, don't rest on your laurels. We want to see some brilliant ideas out of you—otherwise, we're taking the T-shirt away.

Are there robots that feel love?

This question has long been fodder for science fiction. And if it has you dreaming of a day when a robot takes your hand, leans over, and whispers sweet nothings in your ear, you might want to push your internal "pause" button. That day isn't quite ready to dawn.

For now, robots can only react to, mirror, or even intuit our behavior. In Japan, one life-size robot, Kansei, can mimic thirty-six expressions, including those of anger, fear, sadness, happiness, surprise, and disgust. When Kansei hears the word love, it smiles. Japan believes that hastening the development of robots like Kansei is crucial for its society. For instance, these machines can help care for the elderly, who represent one-fifth of the country's population. According to the cadre of artificial intelligence engineers, making robots more effective and omnipresent in society calls for sophisticated versions that can grasp emotions, as well as learn and feel for themselves.

Will these feelings ever include love? British writer David Levy, author of the book *Love + Sex with Robots*, claims that at the very least robots will become more "sexed up" in the coming fifty years. Advanced robot design, he says, will include serving humankind's more carnal desires. As a self-proclaimed augur of robot-human relations, Levy espouses that, like prostitutes or sex dolls, robots will present convincing simulations of love. In addition, he asserts that robots will provide companionship. Even marriage to robots will be possible. This is good news for those of you who haven't had a date since Clinton was in office.

While the notion of a honeymoon in Paris with a robot may seem outlandish, don't be too quick to eschew it. Even though the jury is still out on whether robots will ever love humans, it's been established that humans can feel powerful emotions toward robots, says *New Scientist* magazine.

For instance, U.S. soldiers serving in Iraq felt a unique kinship with Packbots and Talon robots, which dispose of bombs and locate landmines; some soldiers felt profound sadness when these robots were destroyed in explosions. Duke University students name and dress their robotic Roomba vacuums. Some even relate to their Roomba as if it were a family member.

Until the day those Roombas and other robots like them can reciprocate such feelings, we'll need to find love among ourselves. And that seems to be every bit as tricky and elusive as programming the emotion into a machine. Take it from co-medienne Lily Tomlin, who said, "If love is the answer, could you please rephrase the question?"

How does a flak jacket stop a bullet?

"Flak" is an abbreviation of *Fliegerabwehrkanone*, a German word that looks rather silly (as many German words do). There's nothing silly, however, about its meaning: antiaircraft cannon.

Serious development of flak jackets began during World War II, when air force gunners wore nylon vests with steel plates sewn into them as protection against shrapnel. After the war, manufacturers discovered that they could remove the steel plates and instead make the vests out of multiple layers of dense, heavily weaved nylon.

Without the heavy steel plates, the vests became a viable option for ground troops to wear during combat. Anywhere from sixteen to twenty-four layers of this nylon fabric were stitched together into a thick quilt. In the 1960s, DuPont developed Kevlar, a lightweight fiber that is five times stronger than a piece of steel of the same weight. Kevlar was added to flak jackets in 1975.

It seems inconceivable that any cloth could withstand the force of a bullet. The key, however, is in the construction of the fabric. In a flak jacket, the fibers are interlaced to form a super-strong net. The fibers are twisted as they are woven, which adds to their density. Modern flak jackets also incorporate a coating of resin on the fibers and layers of plastic film between the layers of fabric. The result is a series of nets that are designed to bend but not break.

A bullet that hits the outer layers of the vest's material is flattened into a mushroomlike shape. The remaining layers of the vest can then dissipate the misshapen bullet's energy and prevent it from penetrating. The impact of the offending bullet usually leaves a bruise or blunt trauma to internal organs, which is a minor injury compared to the type of devastation a bullet is meant to inflict.

While no body armor is 100 percent impenetrable, flak jackets offer different levels of protection depending on the construction and materials involved. At the higher levels of protection, plates of lightweight steel or special ceramic are still used. But all flak jackets incorporate this netlike fabric as a first line of defense. *Fliegerabwehrkanone*, indeed.

If the Professor on *Gilligan's Island* can make a radio out of a coconut, why can't he fix a hole in a boat?

Those of you under the age of twenty-five might be staring blankly at this page and saying to yourself, "What the heck is *Gilligan's Island*?" Trust us—it was a television show, and a very popular one at that.

Gilligan's Island aired on CBS from 1964 to 1967 (and then *ad infinitum* on TBS, TNT, and Nick at Nite), and was based on a simple premise: A motley group of people on a "three-hour tour" are shipwrecked on a deserted island. One of these shipwrecked tourists was Roy "the Professor" Hinkley, a man with six college degrees and advanced knowledge of technology, science, and obscure island languages. Over the course of ninety-eight episodes, the Professor was able to create radios, lie detectors, telescopes, sewing machines, and other gadgets out of little more than a few coconuts and some bamboo. In short, the Professor was the *MacGyver* of his era. (What? You don't know about the main characters on that show either? We are getting old. Okay, think of *Gilligan's Island* as a goofy precursor to *Lost*.)

With all of Hinkley's technological wizardry, the question must be asked: Why couldn't the Professor fix a simple boat? (Fortunately, most of the male viewers were too concerned with staring at Mary Ann and Ginger to think logically.) The answer, of course, was ratings. Though *Gilligan's Island* was never a smash hit, it was popular enough to last for several seasons. As anybody can tell you, if the Professor was able to fix the boat, there would be no show.

As it turns out, the Professor and his friends had to wait more than a decade to be rescued. Since CBS canceled the 1968 season of *Gilligan's Island* at the last minute, the final episode of the 1967 season found the crew still stranded on its island. In 1978, a special two-part made-for-TV movie, titled *Rescue from Gilligan's Island*, detailed the crew's long-awaited rescue.

It is only fitting that even after a decade, the Professor wasn't able to figure out simple boat repair. Instead, the castaways tied their huts together to make a raft, and they floated to freedom, where they presumably spent the rest of their lives watching reruns of themselves on cable television.

Could humans one day live to be 140?

Yes, and it isn't a far-off possibility. Some scientists believe that within fifty years, people in industrialized nations will routinely live one hundred years or longer. When that time comes, you can bet that a few healthy, energetic individuals will be pushing 140, 145, 150, and beyond.

The average American these days is expected to live seventy-eight years, and the average life expectancy worldwide has been increasing by about two years every decade since the 1840s. Back then, Sweden boasted the population with the most impressive longevity: Healthy folks lived to the ripe old age of forty-five.

The increasing life expectancy is attributed to a number of

factors, such as vaccinations, antibiotics, improved sanitation, and stricter food regulations. Furthermore, improved safety regulations in the workplace and on the road have helped to prevent fatal injuries.

Experts such as James Vaupel, director of the laboratory of survival and longevity at the Max Planck Institute in Germany, believe that life expectancy will continue to climb as techniques improve for preventing, diagnosing, and treating age-related maladies such as heart disease and cancer.

Centennial birthday parties should be commonplace by early in the twenty-second century. That may seem like a long way off, but consider that those centigenarians are being born now. The baby that your sister-in-law just brought home from the hospital may come close to reaching 140.

Without a crystal ball to forecast the medical advances we may achieve, it's impossible to say how long people eventually will be living. But as Daniel Perry, executive director of the Alliance for Aging Research, said, "There is no obvious barrier to living well beyond one hundred."

Healthy, Wealthy, and Weird

How do vaccines work?

If you've ever taken a small child to the doctor for a vaccination, you've probably had to explain that a little bit of hurt—a shot, a finger prick, a dose of sour-tasting medicine—can go a long way toward keeping us well. Then that pesky kid invariably asks, "Why?" You shrug and say, "I don't know. It just does, that's all."

Well, here's a better answer to give: superheroes. Your body has its very own superheroes called T cells and B cells. They're both part of your immune system and kick in whenever a virus shows up. Viruses are tiny microbes that can't live outside a body, so once they get inside you, they want to stay and pretty much take over the place. That's when T cells and B cells go to work and kick some butt.

First, "helper" T cells act as scouts, alerting the rest of your immune system to the presence of the invader and relaying the lowdown on its molecular composition. Then "killer" T cells swing into action, destroying any body cells that have been damaged by the virus. Meanwhile, B cells sweep your blood stream to seek out invaders. When they find a viral microbe, the B cells seize it by releasing special antibodies that lock onto the molecules of the virus, slowing it down and making it possible for large white blood cells called macrophages (the foot soldiers of the immune system) to come and eliminate it. Only one type of antibody will capture any given virus. Once the virus has been defeated, your immune system will remember which antibody it used and will be prepared should that particular unwelcome guest show up again.

What does all of this have to do with vaccines? When you are vaccinated, a doctor injects a tiny bit of a disease into your body. This may consist of a few very weak microbes or an artificial duplicate that researchers produce in a lab. Your body reacts as if the virus was the real thing–it's kind of like having an immune system fire drill. Your B cells manufacture the correct antibodies and add them to your arsenal of immune defenses. In short, you now have superpowers against some dangerous diseases. Pretty cool, huh?

Why don't vaccines work against every disease? Because not all illnesses are brought on by viruses. Some are caused by bacteria and cured by antibiotics. Some are the result of poor nutrition. Others have no known cure. The vaccinations we do have, however, make the world a much healthier place. They have helped many of us live long and prosper. And it doesn't get much more super than that.

Why did doctors perform lobotomies?

Few people have first-hand experience with lobotomized patients. For many of us, any contact with these convalescents comes via Hollywood—that searing image at the end of *One Flew Over the Cuckoo's Nest* of Jack Nicholson, as Randle Patrick McMurphy, lying comatose. Hopefully, we've all experienced enough to know that Hollywood doesn't always tell it like it is. After all, what would be the point of a medical procedure that turns the patient into a vegetable? Then again, perhaps this is the reason that lobotomies have taken a place next to leeches in the Health Care Hall of Shame.

What exactly is a lobotomy? Simply put, it's a surgical procedure that severs the paths of communication between the prefrontal lobe and the rest of the brain. This prefrontal lobe—the part of the brain closest to the forehead—is a structure that appears to have great influence on personality and initiative. So the obvious question is: Who the hell thought it would be a good idea to disconnect it?

It started in 1890, when German researcher Friederich Golz removed portions of his dog's brain. He noticed afterward that the dog was slightly more mellow—and the lobotomy was born. The first lobotomies performed on humans took place in Switzerland two years later. The six patients who were chosen all suffered from schizophrenia, and while some did show post-op improvement, two others died. Apparently this was a time in medicine when an experimental procedure that killed 33 percent of its subjects was considered a success. Despite these grisly results, lobotomies became more commonplace, and one early proponent of the surgery even received a Nobel Prize.

The most notorious practitioner of the lobotomy was American physician Walter Freeman, who performed the procedure on more than three thousand patients—including Rosemary Kennedy, the sister of President John F. Kennedy—from the 1930s to the 1960s. Freeman pioneered a surgical method in which a metal rod (known colloquially as an "ice pick") was inserted into the eye socket, driven up into the brain, and hammered home. This is known as a transorbital lobotomy.

Freeman and other doctors in the United States lobotomized an estimated forty thousand patients before an ethical outcry over the procedure prevailed in the 1950s. Although the mortality rate had improved since the early trials, it turned out that the ratio of success to failure was not much higher: A third of the patients got better, a third stayed the same, and a third became much worse. The practice had generally ceased in the United States by the early 1970s, and it is now illegal in some states.

Lobotomies were performed only on patients with extreme psychological impairments, after no other treatment proved to be successful. The frontal lobe of the brain is involved in reasoning, emotion, and personality, and disconnecting it can have a powerful effect on a person's behavior. Unfortunately, the changes that a lobotomy causes are unpredictable and often negative. Today, there are far more precise and far less destructive ways of affecting the brain through antipsychotic drugs and other pharmaceuticals.

So it's not beyond the realm of possibility that Nicholson's character in *Cuckoo's Nest* could become zombie-like. If the movie gets anything wrong, it's that a person as highly functioning as McMurphy probably wouldn't have been recommended for a lobotomy. The vindictive Nurse Ratched is the one who makes the call, which raises a fundamental moral question: Who is qualified to decide whether someone should have a lobotomy?

Can you get cancer from your cell phone?

This question has come up time and again since 1993, when a man claimed on national television that the radiation from his wife's cell phone caused her to develop cancer. He filed a lawsuit against the manufacturer of the phone, but the case was dismissed due to a lack of evidence.

Seven years later, a neurologist in Baltimore brought a similar suit against his cell phone's manufacturer, alleging that extensive use of his cell phone led to cancer of the brain. Given the magnitude of the case—he sought eight hundred million dollars in damages—it stirred up quite a bit of media attention, and the furor was only fueled by his profession, since it seemed to make his lawsuit all the more credible. Further compounding the issue was an episode of *Larry King Live*, in which an epidemiologist suggested that there might be a link between cell phone use and cancer.

These events were quite sensational, and have continued to re-surface in the media every so often. What frequently is overlooked, however, is that researchers have found no clear connection between cell phone use and cancer. It's certainly true that cellular phones emit small levels of radiofrequency radiation. However, the good news is that the amount is not only miniscule, but also has not been found to cause, advance, or contribute to existing cancerous growths.

Plus, as cell phone technology improves, newer models produce less and less radiofrequency radiation. And if that weren't enough, both the Food and Drug Administration (FDA) and Federal Communications Commission (FCC) limit the amount of radiofrequency radiation that a cell phone can emit—so even if a phone could theoretically produce enough to be harmful, it would never make it to the market.

Although the FCC has safeguards in place regarding radiofrequency regulation, the question of safety comes up often enough to warrant a disclaimer: The FCC does acknowledge that there are potential risks associated with the use of cellular devices, and advises those concerned with radiation to take the simple step of increasing the distance between the body and the source of the radiation. Since it's the cell phone that generates the radiation, one can simply use a headset or a Bluetooth earpiece to keep the phone at a distance from the head (and therefore the brain), thereby reducing exposure.

So go ahead: Breathe a sigh of relief, and go right back to your cell phone conversations. You're not in any real danger of developing cancer as a result, no matter how long you spend with the phone pressed to your ear.

Can you grow your own penicillin?

Mold. Ugh! The slimy stuff covering the leftovers that have been sitting in the fridge for too long. Who needs it? Well, you do if you want to fend off the occasional invasion of deadly bacteria.

Penicillin, one of the world's most powerful antibiotics, is a common form of mold. You've no doubt seen it yourself on bread, potatoes, and other foods. Growing it certainly doesn't take much equipment or skill. Want to try? The British Pharmacological Society recommends the orange method: Pierce a medium-size orange thoroughly with a fork, squeeze it a little to make sure you've gone deep enough to draw out some juice, place it in a shallow dish, and leave it in a cool, dark place at room temperature. After seven to ten days, your orange should sport a fuzzy, bluish-green beard. Penicillin is identified mainly by its color: The blue is *Penicillium italicum;* the green is *Penicillium digitatum.*

Does this mean that applying a piece of moldy fruit to an open wound is a good idea? Could be. For centuries, folk healers employed mold to fight infections. The remedy didn't always work, however, because penicillin spores are elusive little critters. Under natural conditions, it's difficult to find concentrations dense enough to win a face-off with bacteria.

Antibiotic penicillin is derived from a strain called *Penicillium chrysogenum,* which was first isolated in 1943. During World War II, the search for rich sources of penicillin became desperate. Throughout Allied countries, scientists scoured grocery stores for moldy food. Peoria, Illinois, came up with the winner: a rotten cantaloupe that harbored one of the highest concentrations of penicillin ever seen. This "magic cantaloupe" helped the Allies cook up enough penicillin to save the lives of millions of soldiers. It is no exaggeration to say that penicillin did as much to defeat Germany and Japan as bombs.

How exactly does penicillin work its magic? Bacteria multiply via a process known as binary fission. Penicillin contains a substance called beta-lactam that prevents bacteria from reproducing by inhibiting the formation of cytoplasmic membranes, or new cell walls. If they are unable to successfully divide, the bacteria cannot conquer. Cell walls collapse, and the colony rapidly withers away.

Of course, a few individual bacteria will inevitably prove resistant to penicillin, which is why biochemists are always developing alternative antibiotics. If you have a condition that requires frequent antibiotic use, your doctor will probably try to vary the type just to keep those sneaky bacteria on their toes and out of your corpus. Despite this drawback, penicillin remains one of our mightiest medications, providing generations with longer and healthier lives.

Still wondering what to do with of those slimy leftovers? Throw them out. Please! Like your mom said, cleaning out the fridge is a good way to stay healthy, too.

Why do doctors hit your knee with a hammer?

If you're naturally paranoid, you may have considered the possibility that doctors hit your knee just because they can. After all, they could do all sorts of malicious things to us in the name of health, and we would be none the wiser. But thankfully, there's a valid reason for your doctor to whack you on the knee.

The doctor is timing a stretch reflex, a type of involuntary muscle reaction. While you're sitting on a table, the doctor taps a tendon of the quadriceps femoris, the muscle that straightens your leg at the knee. This tendon stretches the muscle suddenly, and sensory neurons send a message to motor neurons in your spinal cord. These motor neurons send a signal to the muscle in your thigh, which contracts. The result is that your leg jerks forward. The reflex is highly efficient—the sensory neurons in your knee are wired directly to the motor neurons in your spinal cord that control the reaction, and the brain isn't even involved.

The body reacts this way to keep you balanced while standing and walking without your having to think about it. Putting weight on the leg as you move or shift your balance causes the muscle to contract to support you. Similar stretch reflexes make the rest of the muscles in your legs and feet do what they're supposed to, as well.

Doctors have been banging on knees to test for spinal cord and nerve disorders for more than a century. A diminished reflex reaction can indicate a serious nerve problem, such as *tabes dorsalis*—the slow degeneration of nerve cells that carry sensory information to the brain. So rest assured—your doctor isn't knocking your knee simply for the entertainment value.

Can the metal plate in Uncle Ed's head get rusty?

Man! If only someone had thought of this potential pitfall during the decades of testing and rigorous Food and Drug Administration scrutiny that it takes to approve surgical implants for use on humans. What a bummer this must be for the orthopedics company that spent millions of dollars bringing prosthetics to market, only to be undone because it overlooked this fundamental aspect of inserting a metal object into a body that is 60 percent water.

Excuse our sarcasm, but goodness gracious, did a fourth-grader come up with this question? No, a metal plate in your Uncle Ed's head—or any other metal surgical implant—won't rust. These implants do, however, pose other problems that we'd love to tell you about.

First, a little bit about surgical metals. There are a number of reasons why an individual would need surgical metal implants, also called biometallics, grafted to the body. A severe cranial injury could mandate a metal plate, but simply breaking an ankle might require a metal screw or rod. Hip-replacement surgery, meanwhile, necessitates a whole network of prosthetics to replace the damaged joint. In general, the field of orthopedics attempts to correct injuries, disorders, or deformities to the skeleton or joints through various means, including surgical implants, casts, splints, and therapy.

The implants are made from a variety of materials that have been thoroughly tested for durability and compatibility with the human body. Two of the most commonly used implant metals are titanium and surgical-grade stainless steel. The titanium alloys used in orthopedics are nearly impervious to corrosion, though they are susceptible to wear and tear. Still, in almost all cases, a surgical implant is rarely removed because of corrosion.

Now, let's discuss the downside of surgical implants. Stainless steel alloys usually contain nickel, which is known to cause allergic reactions in 5 to 20 percent of the population. Titanium, meanwhile, is slightly weaker than stainless steel. There's a chance, albeit a small one, that a titanium implant could crack, causing pain or even releasing chemicals into the body. In addition, depending on the weight and physical activity level of the patient, an implant of any kind could become dislodged or warped over time.

However, continued advancements in the field are helping to make concerns such as these obsolete. The experts have it covered. So, here's a word of advice to whoever thought up this question: Go back to your debate over whether Kirk or Picard is the better *Enterprise* captain, and leave the metal-implant troubleshooting to people with engineering degrees.

Does eating turkey make you tired?

Please put your La-Z-Boy into an upright position—you'll want to be awake for this. While the post-Thanksgiving snooze is a widely experienced phenomenon, turkey is not to blame for it.

The game bird got its bum rap because it contains L-tryptophan, the amino acid popularly known for its sleep-inducing effects. When digested, L-tryptophan travels through the bloodstream to the brain. Once there, it's metabolized into the neurotransmitters serotonin and melatonin. These chemical substances have a calming effect on the body and help us get to sleep.

Now you might think the proof is right there on the holiday platter. But the truth is, scientists know that L-tryptophan can really only make you drowsy if taken purely, on an empty stomach, without any other amino acids. As a protein-rich food, turkey just happens to be loaded with other amino acids.

Furthermore, compared to other common foods, turkey does not contain unusually high levels of L-tryptophan. Chicken, pork, cheese, beef, and soybeans have as much—or more—per equivalent portion.

So what gives? How did turkey get its reputation as a sleeping agent? For one, it's often the centerpiece of the traditional Thanksgiving meal—so much so that we might discount all the starchy, sugary carbohydrates and fats (i.e., stuffing, mashed potatoes, gravy, bread, pumpkin pie, and sweet potatoes smothered with marshmallows) that also take their toll.

Nutritionists say it's not the turkey that makes us tired, but the massive overeating. The average person consumes about three thousand calories and 229 grams of fat—well more than a full day's allotment—over the course of a Thanksgiving meal.

It takes a great deal of energy to digest such a large dinner. And when you have a full stomach, blood is directed away from your brain and nervous system and toward your digestive track. The result? Before you know it, your pants are popped and you're feeling sleepy...very sleepy.

What makes honey so harmful to infants?

Ah, sweet ambrosia of the queen bee! Sugar cane and maple sap aside, honey is perhaps the closest Mother Nature has come to manufacturing candy outright. It's gooey, sticky, and sweet, and it's the only way some people can stomach a cup of tea. Bears love it; in fact, a certain tubby yellow cubby is notorious for the lengths he will go for a "pawful" of the stuff. Rolling around in the mud, pretending to be a little black rain cloud, mooching off his friends: His addiction shows just how tasty honey can be.

But while honey is tasty, it can be very bad for your baby. The sweet stuff is somewhat of a Trojan horse, carrying entire battalions of harmful bacterial spores entrenched within its sticky goodness. These spores produce *Clostridium botulinum* bacteria. Once inside an infant, the bacteria set to work producing a toxin that can lead to infant botulism.

The Centers for Disease Control and Prevention (CDC) is not yet convinced of honey's role in infant botulism, a disease that has also been blamed for Sudden Infant Death Syndrome, or SIDS. There is not enough strong data to warrant a blanket warning regarding honey; however, the CDC hopes parents will look at the evidence for themselves and exercise caution when choosing which foods to give their babies. In the United Kingdom, every jar of honey sold bears a label advising parents against giving it to infants; it's been this way since the connection was made between honey and infant botulism in 1978. Why the CDC has not followed suit in the United States, one can only speculate.

Once a child reaches the twelve-month mark, pediatricians agree it is safe to include honey in your young one's diet. Not only that, but it's healthy and wholesome. Plus, as stated above, it's absolutely delicious. But as far as your baby is concerned, it's best to err on the side of caution.

Why do bruises turn different colors while they're healing?

If you take a lot of beatings, you've no doubt encountered a wondrous rainbow of bruising. Bruises aren't beautiful, but their weird mix of purple, blue, yellow, and even green can be oddly fascinating.

A bruise, or contusion, is an injury in which tiny blood vessels in body tissue are ruptured. As a small amount of blood seeps through the tissue to just below the skin, a deep red or purple bruise forms. The deeper within the tissue the vessels burst, the longer it takes for the blood to reach skin level, and the longer it takes for the bruise to form.

The body is an efficient machine—it's not about to waste the precious iron that's released from the blood when the vessels burst. It dispatches white blood cells to the scene to break down the hemoglobin so the body can salvage the iron.

This chemical breakdown has two notable by-products, each of which has a distinctive color: First, the process produces biliverdin, which is green; then it produces bilirubin, which is yellow. As the deep red hemoglobin, the green biliverdin, and the yellow bilirubin mix, a range of colors results in what we call a bruise. As the body heals, it gradually reabsorbs the by-products, and the skin returns to its normal color.

To minimize bruising, you can apply an ice pack several times a day for a couple of days after you're injured. Or you can invest in some karate classes.

How do smelling salts wake you up?

Smelling salts were found in many homes in the nineteenth century, thanks to the popularity of tight corsets. From time to time, the extreme constriction caused by a fancy lady's corset would reduce the blood supply to the brain, making her "swoon" (preferably into the arms of a dashing gentleman of means). Everyone would gather to enjoy the dramatic gasp as smelling salts snapped the lady back to consciousness.

Smelling salts are a bottled mix of ammonium carbonate and a small amount to perfume. The ammonia compounds naturally decompose in air at room temperature, so when you take the stopper out of the bottle and place it under someone's nose, ammonia gas floods his or her nostrils. If you have ever smelled ammonia gas, you know that it causes you to gasp, which involves inhaling a lot of air. This can jolt someone back to consciousness after he or she has fainted.

While this undoubtedly is entertaining for onlookers, most doctors today say that using smelling salts isn't the ideal way to revive someone who has fainted. The best approach is to simply let the person lie down for five to ten minutes while the body's blood pressure returns to normal. Loosening tight clothing and applying a wet cloth to the forehead are good ideas, too, but above all, you should keep the person calm and still until he or she naturally regains full consciousness.

It's not as dramatic as a sudden gasp and look of horror, but it certainly is more pleasant for the swooner—absent a dashing gentleman of means, of course.

Why does the sun make hair lighter but skin darker?

The key here is a substance called melanin—a bunch of chemicals that combine as a pigment for your skin and hair. In addition to dictating hair and skin colors, melanin protects people from the harmful effects of ultraviolet (UV) light. It does this by converting the energy from UV light to heat, which is relatively harmless. Melanin converts more than 99 percent of this energy, which leaves only a trace amount to mess with your body and cause gnarly problems like skin cancer.

When you head out for a day in the sun and don't put on sunscreen, the sun delivers a massive blast of heat and UV light directly to your skin and hair. The skin reacts to this onslaught by ramping up the production of melanin in order to combat that nasty UV radiation. This is where things get a bit tricky. There are two types of melanin: pheomelanin (which is found in greater abundance in people with lighter skin and hair) and eumelanin (which is found in greater abundance in people with darker skin and hair).

If you're unlucky—that is, if your skin has a lot of pheomelanin—the sun can damage the skin cells, causing a splotchy, reddish sunburn and maybe something worse down the road. After the sunburn, the skin peels to rid itself of all these useless, damaged cells. And then you get blisters and oozing pus, and your skin explodes—no, it ain't pretty. People whose skin has an increased production of eumelanin, on the other hand, are saved from these side effects—the sun simply gives their skin a smooth, dark sheen.

And what's the impact on hair? Well, hair is dead—it's just a clump of protein. By the time hair pokes through the scalp, it doesn't contain any melanin-producing cells. So when the sun damages it by destroying whatever melanin is in it, your mane is pretty much toast—no new melanin can be produced. Consequently, your hair loses its pigment until new, darker strands grow.

The moral of the story? Be sensible in the sun—it's a massive flaming ball of gas, and it doesn't care about your health.

How long can you live without sleep?

Nobody knows for certain, but Dr. Nathaniel Kleitman, the father of modern sleep research, said: "No one ever died of insomnia." Still, what doesn't kill you can still have some nasty side effects.

Various studies have revealed that missing just one night of sleep can lead to memory loss and decreased activity in certain parts of the brain. So if you're planning an all-night cram session for the evening before the big mid-term, you may be better off closing the book and getting a good night's sleep.

Then again, maybe not. Each person's body and brain handle sleep deprivation differently. Some folks are all but useless after one night without shut-eye, while others function normally. It's largely a matter of biology.

Take Tony Wright. In May 2007, the forty-three-year-old British gardener kept himself awake for 226 hours. He said that he was aiming for the world's sleeplessness record and wanted to prove that sleep deprivation does not diminish a person's coherence. Wright admitted to some odd sensory effects during his marathon, but he insisted that his mental faculties were not compromised.

Wright's quest didn't amount to much more than a lot of lost sleep. *Guinness World Records* stopped acknowledging feats of insomnia in 1990 after consulting with experts at the British Association for Counseling and Psychotherapy. The experts believe that sleep deprivation threatens psychological and physical well-being. Muscle spasms, reduced reaction times, loss of motivation, hallucinations, and paranoia can all be triggered by sleep deprivation. That Wright apparently didn't suffer any of these ill effects doesn't mean you won't. Sometimes, it seems, you lose if you don't snooze.

Are tanning beds safer than sun exposure?

Tanning beds may protect naturally pale people from snickers at the beach, but there's no evidence that they guard against the health risks associated with old-fashioned sun worship.

Tanning is a defensive reaction to ultraviolet (UV) radiation, which means it's an indication of skin damage. When the skin sustains damage from UV rays, it produces the brown pigment melanin to help protect against future radiation. Cancer researchers believe that exposure to UV radiation significantly increases the risk of both melanoma and non-melanoma skin cancers. Your body might not effectively repair the resulting damage to its DNA, which can lead to mutations that cause cancer.

So unless your tan comes from a spray bottle, you're increasing your cancer risk. Tanning salons sometimes claim that tanning beds are safer than the sun because they use "moderate" amounts of UVA radiation and even less UVB radiation, which was once thought to be the more dangerous of the two. But research doesn't support this assertion.

In a study published in 2001, researchers at Johns Hopkins School of Medicine and the University of Manchester in England exposed eleven people to ten tanning bed sessions apiece over two weeks. After the sessions were completed, the researchers examined skin and blood samples for signs of molecular change. One such marker is the presence of cyclobutane pyrimidine dimer, which indicates that there has been DNA damage from UV radiation; another is the presence of the p53 protein, which shows that the body is working to repair itself. Both signify the sort of UV damage to cells that can eventually cause cancer, and both were found in all of the participants. The researchers concluded that the damage from a tanning bed session is on par with a day at the beach.

It doesn't appear as if there's such a thing as a safe tan. For the pale-skinned, the choices are to risk cancer, risk turning orange with a spray-on tan, or suck it up and strut onto the beach flaunting pasty flesh that's slathered in sunscreen.

Why do some people dream in black and white?

There's an old saying that nothing is less interesting than another person's dream—unless you are in it. Yet the mystery of dreams has fascinated philosophers and scientists for thousands of years. Aristotle wrote an entire treatise on the subject in 350 BC. Much later, around the dawn of the twentieth century, Sigmund Freud developed an elaborate system of dream interpretation that mostly involved sex.

For all the research that has been done on dream phenomena, surprisingly little has been learned about the function of dreaming. And outside of color symbolists and New Age dream interpreters, few researchers have worked in the area of colors in dreams. But according to studies conducted in the past several years, anywhere from 12 to 20 percent of people dream in black and white.

Several theories have been put forth to explain the drab dream worlds of those 12 to 20 percent. The first of these—corroborated by a number of different dream researchers, including Harvard's J. Allan Hobson, a pioneer on the subject—points to the ephemeral nature of dreams. Everyone has had the experience of waking up from a really cool dream, only to have the details of the plot fade away even as you are trying to confusedly relate them to your bored spouse or roommate at the breakfast table. In the same way, this research suggests, everyone dreams in color, but the memory of the colors fades as quickly as the details do. Most people forget their dreams as soon as they awaken or gradually over the course of a day. But this does not explain why a small percentage of the population reports dreaming in black and white.

Other researchers take a more Jungian tack, suggesting that colors, like the events and objects in dreams, are symbols of the subconscious. Different colors symbolize different emotions—red is passion and drive, blue is calmness and rest, etc.—and shades of gray symbolize a desire to shield oneself from subconscious messages. For example, a red truck in a dream might symbolize passionate assertive-ness, while a gray truck might indicate a desire to mask that assertiveness.

Perhaps the most interesting theory about color in dreams was proposed by University of California–Berkeley psy-chologist Eric Schwitzgebel in a 2002 paper, "Why Did We Think We Dreamed in Black and White?" Schwitzgebel looked at the history of dream research and noticed that the percentage of people who reported colored dreams began to plummet in the late nineteenth century, reached a low of about 30 percent in the late 1950s, and spiked back up as the twentieth century progressed.

Were people really that dull in the first half of the twentieth century? Perhaps. But Schwitzgebel points to something else: The popular forms of media during that period (photography, movies, television) were in black and white. Prior to the invention of photography, black-and-white coloring was rare. (Have you ever seen a classical painting in black and white?) But with the advent of black-and-white photography, many people thought of images—especially everyday images—as being in black and white.

With the rise of film and television, the phenomenon increased. Many of us think of our dreams as movies anyway, and it would be natural for people who had been exposed only to black-and-white film and photography to remember their dreams as drained of color.

If Schwitzgebel's theory is true, one wonders what the future trend might be. Perhaps everyone will start dreaming in CGI.

Why does squinting help you see?

Most of us don't think much about squinting. It's a highly underrated bodily function—it doesn't receive nearly the amount of publicity that, say, belching does. But if we couldn't squint, much of the pageantry of life would elude us. (Okay, that's a bit of an overstatement, but we're trying to get a point across here.)

Light comes into your eyes as individual rays from all directions. The front of each eye has a lens that bends these rays and redirects them onto your retina, which is located at the back of the eye. This focused light forms an image on the retina, in the same way the lens on a camera bends light rays to form an image on film (or on a charge-coupled device in a digital camera).

When everything is working precisely, the lens focuses the rays directly onto the retina, forming crisp images. But for many people, the process works less than ideally. The lens bends some rays in such ways that crisp images form slightly in front of or behind the retina, not directly on it. As a result, things at certain distances appear blurry. This issue is most severe with rays that come into the eye at sharper angles (light from above or below your line of sight), because the lens bends those rays more than the ones that arrive straight on.

So how does squinting help to solve this problem? It covers the top and bottom of your eye, thereby eliminating a lot of the rays that arrive at sharp angles; your lens receives only the head-on rays. Squinting also changes the shape of your eye, in kind of the same way Lasik surgery does. All of this adds up to clearer images.

Basically, squinting acts as a filter—it blocks the peripheral information and allows only the good stuff to enter. So take a moment or two to praise this handy little visual aid. Without squinting, life would be a blurry mess. (Okay, we're exaggerating again, but you get the idea.)

How do you hypnotize someone?

A hypnotist in a show at a Las Vegas casino might employ a flourish of mystical words and fancy moves to mesmerize volunteers, but it's all for show. The real way to hypnotize someone is much simpler and doesn't require a swinging pocket watch or a lovely sequined assistant.

Psychologists don't fully understand hypnosis, but they generally agree that it's a relaxed, hyper-attentive trance, similar to daydreaming. The defining quality of this state is that subjects don't scrutinize and interpret information as they normally would. When a hypnotist makes a suggestion ("Your hands feel heavy, very heavy"), the subject processes the information as if it was real.

A hypnotist induces this trance by getting a subject to relax fully and focus his or her attention on something. This can be as simple as asking the subject to stare at a blank Post-it note for fifteen minutes or so while saying in a soothing voice that the subject is feeling sleepy and his or her eyelids are getting heavy. Generally, the hypnotist reads from a script or recites memorized lines that help the subject to relax different parts of the body and to imagine a carefree state of mind.

Once the subject is in a trance, the hypnotist can make all sorts of suggestions. How exactly does this work? The hypnotist "programs" specific reactions into the subject–

such as, "When I snap my fingers, you will start speaking in gibberish and then you will put a lampshade on your head"– that will be enacted when he or she is summoned from the trance.

Hypnotism doesn't work on everyone. About 10 to 15 percent of the adult population is highly susceptible to it, meaning that these people enter hypnosis easily and respond to many different types of hypnotic suggestions; about 10 percent can't be hypnotized; and the balance of people can be hypnotized but respond in only limited ways to the suggestions. For example, moderately hypnotizable people usually won't follow a suggestion to forget the hypnosis session, whereas the highly hypnotizable will.

For those who are highly suggestible, hypnosis can be an excellent therapeutic tool. It's been effective in reducing nausea from chemotherapy, labor pain, and general anxiety. And if you want to see your friend strut around clucking like a chicken, it can't be beat.

What is face-blindness?

Some people never forget a face. Others can't seem to remember one. We see faces everywhere. Sociologists estimate that an adult who lives in a busy urban area encounters more than a thousand different faces every day. For most of us, picking our friends and loved ones out of a crowd is a snap. Homing in on the faces we know is simply an instinct.

But what if you couldn't recognize faces? Not even the ones that belong to the people you know best? If you seem to spend a lot of time apologizing to your nearest and dearest—saying things like, "Sorry, I didn't see you there yesterday. Did you get a new haircut? Were you wearing a different shirt? A pair of Groucho glasses?"—you might be face-blind.

No, you don't need a new pair of contacts. Face-blind people can have 20/20 vision. And chances are, there's nothing wrong with your memory either. You can be a whiz at Trivial Pursuit, a walking encyclopedia of arcane information, and still not be able to recall the face you see across the breakfast table every morning.

Why? Many scientists believe facial recognition is a highly specialized neurological task. It takes place in an area of the brain known as the *fusiform gyrus*, which is located behind your right ear. People who suffer an injury to this part of their brains are likely to have *prosopagnosia*, a fancy medical term for face-blindness. Others seem to be born that way.

Of course, everybody has occasional problems recognizing faces. For the truly face-blind, however, faces may appear only as a blur or a jumble of features that never quite coalesce into the whole that becomes Bill from accounting or Judy from your softball team.

How many people suffer from face-blindness? Statistics are difficult to come by, simply because many people are not even aware that the inability to recognize faces is a bona fide medical syndrome. However, recent research on random samples of college students indicates that *prosopagnosia* may affect as many as one out of every fifty people, or approximately 2 percent of the population.

What can you do if you think that you are face-blind? Most people with *prosopagnosia* compensate without even knowing it. They unconsciously learn to distinguish people by the way they walk or talk, or perhaps by a distinctive hairdo or article of dress. Many face-blind people write down the information to remind themselves later. Some people who suffer from this affliction compare it to being tone deaf or colorblind—an inconvenient but hardly a life-threatening disability.

As with just about anything, a sense of humor helps, too. Ask your friend to warn you before she frosts her hair or he discards that Pearl Jam T-shirt he's proudly worn since you met in, oh, 1996. And if you really want to make sure that you see your friends and family in a crowd, tell them to wear something that you'll be sure to remember. Maybe the Groucho glasses. They work like a charm every time.

Are human beings still evolving?

Before tackling this one, we need to give you a quick history of human evolution. The creatures that became *homo sapiens* (humans) split from apes between two and four million years ago. A common theory is that everyone started in Africa and slowly spread out across the world. Once isolated from each other, we all evolved slightly differently, which accounts for qualities like different skin colors.

Nowadays, however, we're no longer geographically isolated, so we're unlikely to evolve into an entirely new species. But evolution doesn't just refer to the development of a new species—it's tricky to define. Some people see evolution as a change over time in the gene pool—that's all the genes from all living humans. If that's the case, we *are* still evolving. Each generation has random mutations that may get passed on. These aren't just noticeable changes like albinism or blindness, but also alterations like someone growing up to be a bit taller than usual. Some people have more offspring than others, so more of their genes get passed on, and the gene pool shifts as a result.

Many people prefer to think of evolution as a slow change to the gene pool that is caused by natural selection. Natural selection is a theory that was developed by Charles Darwin: Some random mutations give certain individuals an advantage, and these people survive long enough to have offspring. Maybe the slightly taller person can reach the healthier fruit on the taller trees, for instance. Scientists argue, but the consensus is that we are still evolving according to this theory.

Recent studies have shown that we've evolved in several

ways relatively recently. Researchers at the University of Chicago found two new genes that are involved in developing the brain. One is a version of microcephalin that emerged about 37,000 years ago and is now carried by 70 percent of the global population. The other is a variant of something called the ASPM (abnormal spindle-like microcephaly-associated, if you must know) gene, which arose 5,800 years ago and is now present in about 30 percent of humans.

These two genes coincide with the introduction of new ways of living. The microcephalin gene arrived on the scene in the same period as art, music, sophisticated stone tools, and religious practices. The ASPM variant arose at about the time humans began shifting from hunting and gathering to agriculture and large settlements.

Some people argue that certain races are more evolved than others. But the reality is that everyone has been evolving pretty equally, just in different ways. For instance, Europeans have been found to be more likely to break down lactose (a sugar that is found in milk, which they would have had plenty of from all those cows and goats) than other cultures. On the other hand, several tribes in Northern Africa can break down mannose (a sweet substance found in some plants) more easily, because they've been eating it longer.

So yes, we're evolving, but don't ask scientists what we're evolving into. Evolution is unpredictable.

Everyday Mysteries

Why do wet things look darker than dry things?

We're able to see because light hits objects and bounces off them onto our eyes. The more light that bounces off, the brighter the objects appear. We see colors because objects absorb some frequencies of light but reflect others; the different frequencies appear to us as different colors. Black objects absorb most frequencies and therefore appear dark, while white objects reflect most frequencies and appear bright.

Now that you've had a basic lesson on sight, let's answer the question at hand. We've established that when light hits any object, some of it is reflected back and some is absorbed. Light also can be refracted through an object; this means that it bends and flies off in a different direction, so it doesn't come back to your peepers.

Take a piece of white paper; it reflects most of the light that hits it. If someone were to spill water onto that paper, the water would be absorbed into the paper. Water bends light quite a bit. (Put a stick in clear water and seen how bent it looks.) Since more light is being refracted away from our eyes, we perceive the wet spot to be darker than the rest of the piece of paper.

Water—it may seem kind of boring, but there's a lot more to it than meets the eye.

Why does a compass always point north?

In the early days of navigation, sailors relied upon the position of stars and constellations to determine where they were and where they were going. While this worked reasonably well, the development of the compass brought vast improvements on the navigational front.

So how does a compass work, exactly? An early version of the modern magnetic compass was in use in China as early as AD 850, and the more recognizable mariner's compass was developed in Europe around 1190. There have been many advancements in compass technology since its inception—but in its most basic form, a compass is simply a free-spinning magnet above an image of the four cardinal directions. Since a magnet will naturally point north, all you have to do is let the compass dial swing until it comes to rest and then place the "north" part of the image on the area where the needle is.

Of course, this only works because we know that the needle will always point north. The real question, then, is why does it do this? The truth of the matter is that a magnet doesn't only point north—it points north and south, with each end of the magnet pointing toward either the North or the South Pole of the earth. The earth's magnetic field attracts the ends of the magnet toward the north and the south, but for simplicity's sake, the north-facing end of the needle is all that is denoted on a compass, usually with red paint.

You might be wondering, "How does one know which end is which, without using a compass?" Quite simply, you can use the position of the sun to calibrate the magnet. As long as you remember that the sun rises in the east and sets in the west, you can determine which end of the compass is pointing north.

Why is the sky blue?

What if the sky were some other color? Would a verdant green inspire the same placid happiness that a brilliant blue sky does? Would a pink sky be tedious for everyone except girls under the age of fifteen? What would poets and song-writers make of a sky that was an un-rhymable orange?

We'll never have to answer these questions, thanks to a serendipitous combination of factors: the nature of sunlight, the makeup of Earth's atmosphere, and the sensitivity of our eyes.

If you have seen sunlight pass through a prism, you know that light, which to the naked eye appears to be white, is actually made up of a rainbow-like spectrum of colors: red, orange, yellow, green, blue, and violet. Light energy travels in waves, and each of these colors has its own wavelength. The red end of the spectrum has the longest wavelength, and the violet end has the shortest.

The waves are scattered when they hit particles, and the size of the particles determines which waves get scattered most effectively. As it happens, the particles that make up the nitrogen and oxygen in the atmosphere scatter shorter wavelengths of light much more effectively than longer wavelengths. The violets and the blues in sunlight are scattered most prominently, and reds and oranges are scattered less prominently.

However, since violet waves are shorter than blue waves, it would seem that violet light would be more prolifically scattered by the atmosphere. So why isn't the sky violet? Because there are variations among colors that make up the spectrum of sunlight—there isn't as much violet as there is blue. And because our eyes are more sensitive to blue light than to violet light, blue is easier for our eyes to detect.

That's why, to us, the sky is blue. And we wouldn't want it any other way.

What happens when you drop a penny from a tall building?

If you've ever taken the elevator to the observation deck on the Empire State Building, you may have contemplated the ultimate fate of coins tossed from that awesome height onto the street hundreds of feet below. Would they leave small, penny-size craters as they embed themselves into the tarmac? If one hit a person, would it kill him or her instantly? Surely a penny falling such a tremendous distance would reach a speed similar to that of a bullet and strike with the same deadly effect.

To determine the force of impact, we need to know the mass and speed of the penny. The typical modern penny weighs 2.5 grams, so we'll use that number. Calculating the speed is a bit more complicated.

Terminal velocity is the maximum speed that something will reach when freefalling through the atmosphere. Falling objects don't just continue to drop faster and faster. As gravity pulls down on the object, friction and wind resistance (drag) work to slow the object down. The faster the speed, the greater the drag. At a certain point, the speed is so fast that the drag is just as strong as gravity, so the object stops accelerating and continues falling at the same speed.

Now, it's a not-so-simple matter of solving the equations for drag and terminal velocity. It's not so simple because some of the factors in those equations can vary depending on the circumstances. Is the penny old and rough, or is it polished to a shine? These factors will change the drag of air passing over the penny's surface. Is the skyscraper in a city at a high altitude, such as Denver, where the air is thinner? Thinner air offers less drag. The terminal velocity of a penny is between thirty and forty-five miles per hour.

Let's say we have a 2.5-gram penny striking at about thirty-five miles per hour. Let's compare this to a bullet. Some bullets travel more than a thousand miles per hour, and even the slowest go at least three hundred miles per hour. Bullets weigh 125 grams or more. Considering that force equals mass times acceleration (in this case, negative acceleration, since the penny and the bullet strike something and come to a stop), you don't even need to do the math to see that a penny at terminal velocity doesn't strike with anything near the force of a bullet. Someone could be harmed by a dropped penny if it hit him or her right in the eye, but otherwise, it wouldn't do any real damage.

Why does mercury rise or fall depending on the temperature?

It's a standard cartoon gag: The sun is beating down, eggs are frying on the sidewalk, and people are fanning themselves in overheated agony. Cut to a shot of the thermometer; mercury strains against its glass confines before blowing out the top like water from a geyser. It's exaggerated, but not as much as you may think.

Mercury expands as the temperature increases; as the temperature decreases, it contracts. (If this seems counterintuitive, it's because water reacts to temperature in the opposite way, expanding as it freezes.) As mercury expands, it is channeled upward through the thermometer's thin, hollow center; it will expand as long as the temperature continues to rise, which is why most thermometers have a reservoir at the end.

Daniel Gabriel Fahrenheit—a German glassblower, physicist, and engineer—invented the mercury thermometer in 1714. Fahrenheit chose mercury to gauge changes in temperature for two reasons: The element is liquid at room temperature, and it expands evenly as the temperature rises. Unfortunately, mercury is also quite dangerous—even its vapors are poisonous. That's why, for the most part, digital thermometers have replaced those that are filled with mercury—there always is a chance that glass thermometers will break.

So back to the time-honored cartoon gag, from which you can learn the following lesson: If you notice eggs frying on the sidewalk, be wary of any mercury thermometers you see. Each is like a loaded gun, just waiting for the temperature to get hot enough to make it blow like a toxic Old Faithful.

Can people get sucked out of airplanes?

Unfortunately, yes. All sorts of objects, including human beings, can be hurled out of shattered windows, broken doors, and other holes in the skin of an aircraft. The problem in any case like this is explosive decompression—a situation in which the keys to survival include the height at which the plane is flying and the size of the aircraft cabin itself.

Most passenger planes are pressurized to approximate an altitude of eight thousand feet or less. If there is a break in the skin of the plane at an altitude considerably higher than that—commercial jets normally fly at thirty thousand feet or more—all sorts of bad things can happen. And quickly. For one thing, a lack of oxygen will render most people unconscious in little more than a minute at thirty-five thousand feet.

As for being sucked out of the plane, this sort of tragedy generally occurs when the decompression is very sudden. The difference in pressure between the inside and outside of the plane causes objects to be pulled toward the opening. Whether or not someone survives sudden decompression depends on several things, including luck.

Critical care nurse Chris Fogg was almost sucked out of a medical evacuation plane on a flight from Twin Falls, Idaho, to Seattle in 2007. Fogg had not yet buckled his seat belt when a window exploded while the plane was flying at approximately twenty thousand feet. His head and right arm were pulled out of the window, but Fogg held himself inside the aircraft with his left hand on the ceiling and his knees jammed against a wall.

Fogg, who weighed 220 pounds, summoned enough strength to push himself backward, which allowed air to flow between his chest and the window. This broke the seal that had wedged him in the opening. The pilot managed to get the plane to a lower altitude, and everyone aboard—including a patient who had been hooked to an oxygen device—survived the ordeal.

Not everyone has been so fortunate in cases of explosive decompression. In 1989, a lower cargo door on a United Airlines flight came loose at twenty-three thousand feet, and the loss of pressure tore a hole in the cabin. Nine passengers were sucked out of the plane, along with their seats and the carpeting around them. A year earlier, an eighteen-foot portion of roof tore off an Aloha Airlines flight at twenty-four thousand feet, hurling a flight attendant out into the sky.

Although scenes of people and debris whistling all over the place in adventure movies have been exaggerated, the threat of explosive decompression is real—even if it is rare. It's a good idea to take the crew's advice about wearing your seat belt at all times, and if there is a sudden loss of oxygen, put on that mask in a hurry. They're not kidding about the possible consequences.

What is the speed of dark?

Most of us believe that nothing is faster than the speed of light. In high school physics, we learned that something traveling faster than light speed could theoretically go back in time. This would allow for the possibility that you could go back in time and kill your grandfather and, thus, negate your existence—a scenario known as the Grandfather Paradox. Or more horrifyingly, you could go back in time in order to set up your future parents as you skateboard around to the musical stylings of Huey Lewis and the News.

Yet there is something that may be faster than the speed of light: the speed of dark. Or maybe not. The speed of dark may not even exist. When you're talking about astrophysics and quantum mechanics, nothing is certain (indeed, uncertainty might be said to be the defining principle of modern physics).

Observations and experiments in recent years have helped astrophysicists shape a more comprehensive understanding of how the universe operates, but even the most brilliant scientists are operating largely on guesswork. To understand how the speed of dark theoretically might—or might not—exceed the speed of light, we'll have to get into some concepts that are usually reserved for late-night bong sessions.

As with much of astronomy, our explanation is rooted in the Big Bang. For those of you who slept through science class or were raised in the Bible Belt, the Big Bang is the prevailing scientific explanation for the creation of the universe. According to the Big Bang theory, the universe started as a pinpoint of dense, hot matter. About fourteen billion years ago, this infinitely dense point exploded, sending the foundations of the universe into the outer reaches of space.

The momentum from this initial explosion caused the universe to expand as it drove the boundaries outward. For most of the twentieth century, the prevailing thought was that the rate of expansion was slowing down and would eventually grind to a halt. Seemed logical enough, right?

In 1998, however, astronomers participating in two top-secret-sounding research projects, the Supernova Cosmology Project and the High-Z Supernova Search, made a surprising discovery while observing supernovae events (exploding stars) in the distant reaches of space. Supernovae are handy for astronomers because just prior to exploding, these stars reach a uniform brightness. Why is this important? The stars provide a standard variable, allowing scientists to infer other statistics, such as how far the stars are from Earth. Once scientists know a star's distance from Earth, they can use another phenomenon known as a redshift (a visual analogue to the Doppler effect in which light appears differently to the observer because an object is moving away from him or her) to determine how much the universe has expanded since the explosion.

Still with us? Now, based on what scientists had previously believed, supernovae should have appeared brighter than what the redshift indicated. But to the scientists' amazement, the supernovae appeared dimmer, indicating that the expansion of the universe is speeding up, not slowing down. How could this be? And if the expansion is quickening, what is it that's driving it forward and filling up that empty space?

Initially nobody had any real idea. But after much discussion, theorists came up with the idea of dark energy. What is dark energy? Ultimately, it's a made-up term for the inexplicable and incomprehensible emptiness of deep space. For the purposes of our question, however, one important attribute is that dark energy is far faster than the speed of light—it's so fast, in fact, that it is moving too quickly for new stars to form in the empty space. No, it doesn't make a whole heck of a lot of sense to us either, but rest assured, a lot of very nerdy people have spent a long time studying it.

Of course, there may be a far simpler answer, one posited by science-fiction writer Terry Pratchett: The speed of dark **must** be faster than the speed of light—otherwise, how would dark be able to get out of the way?

Does running through the rain keep you drier than walking?

It makes intuitive sense that running through the rain will keep you drier than walking. You will spend less time in the rain, after all. But there's a pervasive old wives' tale that says it won't do any good. So every time there's a downpour and you need to get to your car, you are faced with this confounding question: Should you walk or run?

The argument against running is that more drops hit your chest and legs when you're moving at a quicker pace. If you're walking, the theory goes, the drops are mainly hitting your head. So the proponents of walking say that running exposes you to more drops, not fewer.

Several scientists have pondered this possibility (after finishing up their actual work for the day, we hope). In 1987, an Italian physicist determined that sprinting keeps you drier than walking, but only by about 10 percent, which might not be worth the effort and the risk of slipping. In 1995, a British researcher concluded that the increased front-drenching of running effectively cancels out the reduced rain exposure.

These findings didn't seem right to two climatologists at the National Climatic Data Center in Asheville, North Carolina, so they decided to put them to the test. In 1996, they put on identical outfits with plastic bags underneath to keep moisture from seeping out of the clothes and to keep their own sweat from adding to the drenching. One person ran about 330 feet in the rain; the other walked the same distance. They weighed the wet clothes, compared the weights to those when the clothes were dry, and determined that the climatologist who walked got 40 percent wetter than the one who ran.

In other words, run to your car. You're justified—no matter how silly you might look.

How do we know what's in Earth's core?

The moles aren't talking, so we've had to figure it out the hard way. Geologists say that the center of Earth is a massive metal core. The inner section of the core, which is about 1,500 miles in diameter, is mostly iron and contains some nickel. Surrounding the inner section is a 1,400-mile-thick layer of liquid iron and nickel called the outer core. This core is covered by an extremely hot, slow-moving liquid called the mantle that is 1,800 miles thick; the mantle accounts for the bulk of the planet's mass. Above the mantle is the outer crust, five to thirty miles thick and made up of cool rock, on top of which is where we live and play.

We are familiar with the crust, and we sometimes see evidence of the mantle from volcanic eruptions. Everything we know about the areas below the mantle comes from guesswork and clever remote measurements.

The most useful measurement device is the earthquake. Vibrations from an earthquake generate seismic waves that not only move across the surface of the crust, but also through the planet's interior. Just like light waves, seismic waves change speeds as they pass through different types of material. One effect of changing speeds is that the waves refract (turn) at the boundaries between two layers, just as light refracts at the boundary between air and water or as it goes through a lens. Earthquakes produce two types of waves—P waves and S waves—that move through material distinctly and provide seismologists with lots of data.

By noting the time it takes for waves to travel through the planet and observing the patterns of these waves, seismologists have estimated the general densities and locations of different layers of material. The most striking piece of data is a massive "shadow zone" of S waves. Essentially, something in Earth's core blocks S waves that are generated on one side of the planet from reaching the opposite side of the planet.

This suggests that part of Earth's core is liquid, since S waves can move through solid material but not liquid. P waves can move through liquid; their patterns indicate that they encounter an inner solid core after the liquid layer. The intense pressure at the center of the planet apparently prevents the inner core from liquefying.

There's one final piece of the puzzle. All of the planets came from the same swirling mass of matter that formed the solar system. Based on the composition of meteorites left over from this original space junk, scientists have determined the general mix of elements that would have gone into forming Earth. This analysis shows that the planet should include a huge amount of iron, which isn't accounted for in Earth's crust, atmosphere, or mantle. So it must be in the core. As a young Earth was cooling, the heavy iron presumably sank to the center. If only the moles could tell us for certain.

Does sound travel quicker through water or air?

When you came across this question in the table of contents, you probably thought, "Everyone knows the answer to that one. It's **so** obvious." This, of course, is exactly why we chose to include it. Once again, it is our honored duty to inform you that you are 100 percent wrong. Sound actually travels much faster through water than air. We'll get to the hows and whys in a moment, but let's start with the reason that you and everyone who is not an engineer were so sure that you knew the correct answer.

The confusion probably stems from the fact that we humans are designed to process sound waves that are transmitted through air, not water. Perhaps you learned this as a child at the local swimming pool, when you and your buddy dipped your heads beneath the surface and then attempted to communicate. Although you may have heard **something**, it was most likely an unintelligible gurgle.

The act of turning vibrations into sounds involves a series of bones in the inner ear—the malleus, incus, and stapes—and bone conductivity is 40 percent less effective underwater. Furthermore, when the auditory canal fills with water, the eardrum (another major aspect involved in the sensation of hearing) doesn't vibrate properly. As a result, many people just assume that sound travels less efficiently through water.

But the truth is, sound travels approximately five times faster through water than air. We say "approximately" be- cause there are several variables to consider, such as tem- perature, altitude, and humidity. But whether you are talking about cold air or warm water, there must be some sort of medium in order to facilitate the transmission of sound waves. (There is no sound in the vacuum of space.)

Essentially, sound waves are just molecules bouncing off of other molecules until they reach your ears and are pro- cessed as sound. Because water is denser than air, these molecules are packed closer together and sound waves are able to travel at a greater velocity. For the same reason, sound can travel farther through water than air before dis- sipating. (Sound, in fact, travels fastest through solids such as metal. That's why people have been known to put an ear to a rail to hear if a train is approaching.)

Gases such as air are relatively poor conductors of sound—they just happen to be the right environment for the human auditory system. So, yes, we kind of set you up with this one—and for that, we apologize. It was all in the name of providing you with a nugget of valuable knowledge.

How far do you have to dive underwater to escape gunfire?

Unlike outrunning an explosion, this action-hero escape plan actually works. A 2005 episode of the Discovery Channel's *MythBusters* proved that bullets fired into the water at an angle will slow to a safe speed at fewer than four feet below the surface. In fact, bullets from some high-powered guns in this test basically disintegrated on the water's surface.

It might seem counterintuitive that speeding bullets don't penetrate water as easily as something slow, like a diving human being or a falling anchor. But it makes sense. Water has considerable mass, so when anything hits it, it pushes back. The force of the impact is equal to the change in momentum (momentum is velocity multiplied by mass) divided by the time taken to change the momentum.

In other words, the faster the object is going, the more its momentum will change when it hits, and the greater the force of impact will be. For the same reason that a car suffers more damage in a head-on collision with a wall at fifty miles per hour than at five miles per hour, a speeding bullet takes a bigger hit than something that is moving more slowly.

The initial impact slows the bullet considerably, and the drag that's created when it moves through water brings it to a stop. The impact on faster-moving bullets is even greater, so they are more likely to break apart or slow to a safe speed within the first few feet of water.

The worst-case scenario is if someone fires a low-powered gun at you straight down into the water. In the *MythBusters* episode, one of the tests involved firing a nine-millimeter pistol directly down into a block of underwater ballistics gel. Eight feet below the surface seemed to be the safe distance—the ballistics gel showed that the impact from the bullet wouldn't have been fatal at this depth. But if a shot from the same gun were fired at a thirty-degree angle (which would be a lot more likely if you were fleeing from shooters on shore), you'd be safe at just four feet down.

The problem with this escape plan is that you have to pop up sooner or later to breathe, and the shooter on shore will be ready. But if you are a proper action hero, you can hold your breath for at least ten minutes, which is plenty of time to swim to your top-secret submarine car.

Why does metal spark in a microwave?

Microwave ovens work by permeating your food with microwave radiation. That sounds a little scary, but don't worry: We're not talking about the kind of radiation that gave us the giant lizard that stomped Japan. Instead, this radiation excites the water molecules that make up a large portion of every kind of food we eat. The vibrating water molecules start to get hot, which in turn heats the food.

Simple so far, right? It gets a little trickier. Metal responds quite differently to the electromagnetic field that a microwave oven creates. Unlike water, which can absorb the microwave energy, metal *reflects* the radiation. And the energy of the electromagnetic field can also cause a charge to build up in metal—especially if the metal is thin and pointy, like the tines of a fork, the handle of a Chinese take-out box, or the decorative rim on your Young Elvis commemorative plate.

When enough of a charge builds up, all of that energy in the metal can leap joyfully through the air. We see this leap as a spark—like a small-scale bolt of lightning. These arcs of electricity are most likely to emanate from sharp edges, like the tines of a fork or the ridges of crumpled aluminum foil. A solid object with no sharp edges should be okay, because any electrical charge that develops is more likely to spread itself around evenly.

But even then, there's a danger—the metal could reflect the microwave radiation back at the magnetron tube that creates the electromagnetic field. This could damage the magnetron tube, and then you'd be stuck with a useless microwave.

So here's an equation for you: metal + microwave = really bad idea. Stick to food.

How do fireworks form different shapes?

Fireworks have been delighting people (and on the negative side, blowing off fingers) for more than seven hundred years, and the design hasn't changed much in that time. Getting those fireworks to form complex shapes is one of the trickier challenges, but the basic idea is still fairly old school.

To understand what's involved, it helps to know some fireworks basics. A fireworks shell is a heavy paper container that holds three sections of explosives. The first section is the "lift charge," a packet of black powder (a mixture of potassium nitrate, sulphur, and charcoal) at the bottom of the shell. To prepare the shell for launch, a pyrotechnician places the shell in a mortar (a tube that has the same diameter as the shell), with the lift charge facing downward. A quick-burning fuse runs from the lift charge to the top of the mortar. To fire the shell, an electric trigger lights the quick fuse. It burns down to ignite the black powder at the bottom of the shell, and the resulting explosion propels the shell out of the mortar and high into the air.

The second explosive section is the "bursting charge," a packet of black powder in the middle of the shell. When the electric trigger lights the quick-burning fuse, it also lights a time-delay fuse that runs to the bursting charge. As the shell is hurtling through the air, the time-delay fuse is burning down. Around the time the shell reaches its highest point, the fuse burns down to the bursting charge, and the black powder explodes.

Expanding black powder isn't exactly breathtaking to watch. The vibrant colors you see come from the third section of explosives, known as the "stars." Stars are simply solid clumps of explosive metals that emit colored light when they burn. For example, burning copper salts emit blue light and burning barium nitrate emits green light. The expanding black powder ignites the stars and propels them outward, creating colored streaks in the sky.

The shape of the explosion depends on how the manufacturer positions the stars in the shell. To make a simple ring, it places the stars in a ring around the bursting charge; to make a heart, it positions the stars in a heart shape. Manufacturers can make more complex fireworks patterns, such as a smiley face, by combining multiple compartments with separate bursting charges and stars in a single shell. As the fuse burns, these different "breaks" go off in sequence. In a smiley face shell, the first break that explodes makes a ring, the second creates two dots for the eyes, and the third forms a crescent shape for the mouth.

It's hard to produce designs that are much more complex than that, since only a few breaks can be set off in quick succession. So if you're hoping to see a fireworks tribute to origami, you're out of luck.

Why does salt melt ice?

The link between salt and hypertension is clear. Yet each winter highway departments dump billions of tons of salt onto the nation's frozen highways, with complete disregard for the health of the nation's transportation arteries. Is it any wonder that our highways are pitted with potholes? More importantly, has anybody considered the link between salt on the highways and the increase in incidents of road rage?

Until somebody tackles these really important questions, we'll have to satisfy ourselves with the explanation for why all that salt gets dumped in the first place. No, it's not to add a savory zing to the roads—it's to melt the ice. But in order to understand how salt melts ice, we'll need to take a trip back to chemistry class, where we learned how water freezes.

Water, as we all know, is known chemically as H_2O—it's two hydrogen atoms and one oxygen atom that bind together to form a molecule. These molecules are always bouncing about, though this movement is contingent on temperature. Heat speeds up the movement of molecules; cold slows it down. Eventually, if the temperature gets cold enough, water molecules cling together to form ice. As even we chemistry dummies know, this happens at thirty-two degrees Fahrenheit.

In order to melt ice, either the outside temperature needs to go up (a sunny day, for example) or the freezing point of the water itself must be lowered to below thirty-two degrees. Enter salt. When salt is dumped onto ice, the salt molecules bind with the water molecules; as a result, a colder temperature is required for the salt and water molecules to break apart so that ice can be formed.

However, this process requires that some water be present—if the temperature is too cold, there won't be any liquid water molecules to which the salt molecules can bind. (This is why in especially cold climates, highway trucks sometimes dump sand onto roadways—sand doesn't bind with water, but it does provide better traction.)

This principle is applied in other fields as well. In culinary school, chefs are taught that adding salt to water raises the temperature at which water boils, and the same idea is behind how antifreeze keeps your engine from freezing or locking up in the winter.

It's tempting to think that this would work with the human body as well: A couple of teaspoons of salt should warm you up, right? We don't recommend it. Though salt may theoretically warm your heart, it will probably wear it out, too.

Why do old newspapers turn yellow?

The daily newspaper is an amazing achievement of labor and technology that we sometimes take for granted. It has been said that any one issue of the Sunday *New York Times* contains more information than an educated person in the eighteenth century consumed in a lifetime.

You probably wouldn't want to try to make it through the Sunday *Times* during your morning commute, but this gives you some idea of just how much information even your garden variety daily newspaper offers. And what do you pay for the daily delivery of such a wealth of knowledge? Somewhere in the neighborhood of seventy-five cents per issue. Not a bad deal.

In order to bring you all that content—international reports, the latest from Washington, crime news, arts reviews, horoscopes, sports scores and stats, and that columnist who irritates you so much but whose articles you never fail to read—publishers need to keep costs in line. That's why newsprint is used: It's cheap.

Like all paper, newsprint is made from wood, which is composed primarily of cellulose, hemicellulose, and lignin. Cellulose, which is kind of like the flesh of a tree, is white. Lignin serves as a tree's bones, giving it strength and stability. Even though lignin is strong, it deteriorates with exposure to oxygen. Consequently, paper with lignin in it becomes yellow and brittle over time.

When high-quality paper is manufactured, the lignin is removed. To produce newsprint as cheaply as possible, the manufacturer skips the lignin-removal process. This leaves newsprint especially vulnerable to the elements, so it deteriorates more quickly than more luxurious paper stock. The best way to save newspaper articles is to make laser copies of them on high-quality paper that will last for years without becoming yellow or brittle.

Or, if you are cheap, you can keep the original newsprint and slow the deterioration process by keeping it protected from light and oxygen. In this case, if you want to admire your collection of historic news, you'll need to master the art of reading in the dark while holding your breath.

Why does a boomerang come back?

Because if it didn't, it would just be a stick. Seriously, a boomerang flies in a circle because its simple structure—essentially two wings joined together—combines basic physical forces in a unique way.

We won't beat you over the head with details of wing physics; they could fill a book. The important thing to understand is that a wing diverts the flow of the air rushing past it, which generates lift. In other words, as a wing zips along through air, the air follows the wing's curved shape and leaves the back of the wing moving downward.

This lifts a wing in two ways: The air pushing downward causes an equal and opposite reaction in the wing, which forces the wing upward. At the same time, the moving air creates a disturbance that drives air away from the space above the wing, which creates a drop in air pressure that essentially sucks the wing upward.

The correct way to throw a boomerang—they come in right- and left-handed versions, by the way—is nearly vertically. This allows lift from the two wings to push sideways rather than upward. As the boomerang spins, whichever wing is at top at any particular moment generates more lift than the wing that's at the bottom. Why? Because the top wing is moving through the air more quickly than the bottom one. (Note that if the boomerang were spinning while sitting on a rod, like a helicopter rotor, the same amount of air would be moving past both wings all the time. But because you throw the boomerang, you add forward motion to the mix.)

At the top of the rotation, the wing is spinning forward, in the same direction as the throw. At the bottom of the rota- tion, the wing is spinning backward, opposite the forward direction of the throw. Since whichever wing is on top is moving forward more quickly (forward motion of the spin plus forward motion of the throw), more air rushes past, which generates more lift.

This extra lift pushes on each wing at the top the spin. However—and this is where things get weird—there's a delayed reaction, and the push doesn't take effect until the wing rotates another ninety degrees. This weird phenomenon of spinning objects is called gyroscopic procession. It's the same thing that makes a bicycle turn when you take your hands off the handles and lean to the left or right: You're pushing on the top of the wheel, but the front of the wheel turns.

As the spinning boomerang flies, lift is constantly pushing sideways on its leading edge. The push gradually turns the boomerang; this makes it travel in a circular path, right back to the thrower. At the end of the flight, the force of the lift has pushed the boomerang on its side, which makes it easier to catch between your hands, like a Frisbee.

Where is the dirtiest place in the house?

Believe it or not, there's a place in the house that's even dirtier than the toilet or the trash can. We're talking about the kitchen sink.

Say what? How can the place where you wash things get so scuzzy? Well, most people use the kitchen sink to rinse and prepare items like chicken carcasses and store-bought fruits and vegetables.

According to Dr. Philip Tierno, director of New York University's microbiology department and the author of *The Secret Life of Germs*, these raw foods carry tons of potential pathogens, including salmonella, campylobacter, and E. coli. And you're just splish-splashing and spreading them around.

Right now, your kitchen sink's faucet handles and basin are probably teeming with microscopic creepy-crawlies. In fact, there are typically more than five hundred thousand bacteria per square inch in the sink drain alone! The average garbage bin has only about 411 bacteria per square inch. Grossed out yet? Well, there's more.

What about those damp dishrags and sponges that you toss into the sink? You know, the ones that you use to wipe down the kitchen counter? Tierno says that they can hold "literally billions of bacteria." Dr. Charles Gerba (a.k.a. Dr. Germ), an environmental microbiology professor at the University of Arizona, suggests that it might be better to live like a slobby bachelor than to meticulously clean kitchen surfaces with a salmonella-soaked sink sponge. According to Gerba's findings, your kitchen sink is dirtier than a post-flush toilet bowl. Go ahead, take a moment to gag.

Done gagging? Now, grab your scrub brush. You can have a much cleaner kitchen sink in a jiffy. Just mix one tablespoon of chlorine bleach with one quart of water, and use this solution to scrub your sink basin clean. Tierno recommends that you do this twice a week. In between, clean your kitchen sink and counter with an antibacterial product every time you prepare or rinse food.

As for sanitizing those cross-contaminated kitchen sponges, throw them in the washing machine with bleach, run them through the dishwasher's dryer cycle, or nuke them for one minute at high power in the microwave. Even then, don't use a sponge for more than a month. Or you can do away with kitchen sponges altogether. Wipe up food spills with paper towels and dump everything into the trash–where dirty stuff belongs.

How does glow-in-the-dark stuff work?

Switch off the light in the bedroom of a typical eight-year-old and you might see a ceiling glowing with stars, planets, and dinosaurs. "'Tis vile witchcraft!" you may shout, reaching around for some rope and a torch. Wait! Before you burn the kid at the stake, let's review simple physics.

Atoms gain and lose energy through the movement of electrons. Electrons are the tiny, negatively charged particles that orbit the atom's positively charged nucleus. When something energizes an atom, an electron jumps from a lower orbital (closer to the atom's nucleus) to a higher orbital (farther from the atom's nucleus). Basically, the atom is storing energy that will be released in some form when the electron falls back to the lower orbital.

Light is one thing that can energize an atom in this way. When a light photon (an individual packet of light energy) hits the atom, that energy boosts an electron to a higher orbital. Some atoms can release energy as light: When the electron falls back to a lower orbital, the stored extra energy is emitted as a light photon.

Glow-in-the-dark stuff contains atoms that do just this; they're called phosphors. When you turn on the lights in your kid's room, the photons from the light excite these atoms and boost their electrons to higher orbitals. When you turn off the lights, the atoms release the stored energy. The electrons return to a lower level, emit photons, and the atoms glow. There's some energy loss in the process, so the glowing light of the little dinosaur will be of a different frequency (a different color) than the light that excited the atom in the first place.

Phosphors are everywhere. Your fingernails, teeth, and bodily fluids all contain natural phosphors. Your white clothes are phosphorescent too, thanks to whitening agents in laundry detergent. This is why all this stuff glows under a black light. The black light emits invisible ultraviolet light, which causes the phosphors to glow. (Dark clothes may contain the phosphor, too, but the dark pigments absorb the UV light.)

Most phosphors have very short persistence—the atoms release the light energy immediately after they're charged, so they glow only when light is shining on them. By contrast, glow-in-the-dark stickers and the like are made of phosphors such as zinc sulfide and strontium aluminate that have unusually long persistence, so they keep glowing after you turn out the lights. Manufacturers mix these phosphors with plastic to make glow-in-the dark items in many shapes and sizes.

Other types of phosphors react to radiation from radioactive elements and compounds rather than from visible light radiation. This is how the hands on some watches glow with no charging required: They're coated with a radioactive isotope of hydrogen called tritium or promethium that's mixed with phosphors.

As you now know, there's a scientific explanation for glowing stars on the bedroom ceiling. So you can put away the rope and the torch.

How did people wake up before there were alarm clocks?

Everyone has a trick for waking up on time. Some people put the alarm clock across the room so that they have to get out of bed to turn it off; some set the clock ahead by ten or fifteen minutes to try to fool themselves into thinking that it's later than it is; some set multiple alarms; and some—those boring Goody Two-Shoes types—simply go to bed at a reasonable hour and get enough sleep.

We don't necessarily rely on it every day, and some of us definitely don't obey it very often, but just about everybody has an alarm clock. How did people ever wake up before these modern marvels existed?

Many of the tough problems in life have a common solution: hire someone else to do it. Long ago in England, you could hire a guy to come by each morning and, using a long pole, knock on your bedroom window to wake you up so that you would get to work on time. This practice began during the Industrial Revolution of the late eighteenth century, when getting to work on time was a new and innovative idea. (In the grand tradition of British terminology that makes Americans snicker, the pole operator was known as a "knocker-up.") There's no word on how said pole operator managed to get himself up on time, but we can guess.

The truth is, you don't need any type of alarm, and you never did. Or so science tells us. Your body's circadian rhythms give you a sort of natural wake-up call via your body temperature's daily fluctuation. It rises every morning regardless of when you went to bed. Studies conducted at Harvard University seem to indicate that this rising temperature wakes us up (if the alarm hasn't already gone off).

Another study, conducted at the University of Lubeck in Germany, found that people have an innate ability to wake themselves up very early if they anticipate it beforehand. One night, the researchers told fifteen subjects that they would be awakened at 6:00 AM. Around 4:30 AM, the researchers noticed that the subjects began to experience a rise in the stress hormone adrenocorticotropin. On the other two nights, the subjects were told that they would get a 9:00 AM wake-up call—but those diligent scientists shook them out of bed three hours early, at 6:00 AM. And this time, the adrenocorticotropin levels of the subjects held steady in the early morning hours.

It seems, then, that humans relied on their bodies to rouse them from the dream world long before a knocker-up or an alarm clock ever existed.

Does plastic ever decompose?

Those tree-hugging environmentalists are always trying to make you feel bad about carrying your groceries home in plastic bags. "Plastic takes five hundred years to decompose," they say. But really, how do they know? Plastics weren't invented until the twentieth century. It's not like Copernicus toted his astronomy books around in a plastic bag that is now blowing in the wind somewhere in Poland.

First, it's important to understand that there are many different types of plastic. Plastics are made from petroleum and other fossil fuels, and can be manufactured to various strengths and resistances to hot and cold. For instance, polythene is the type of plastic used to make children's toys, food packaging, and plastic shopping bags; it's not particularly strong. At the other end of the spectrum are epoxies, which are used, among other things, to make airplanes, including Harrier jets. Naturally, the stronger, more durable plastics take longer to decompose.

Let's stick with plastic shopping bags, a main eco-offender cited by environmentalists. In fairness, the claims of five hundred or even one thousand years are not completely unfounded. Bacteria and other microorganisms cannot feast on plastic bags the way they do on, say, banana peels because they do not recognize the manufactured substance as food. As a result, plastic bags do not biodegrade.

Plastic bags will, however, photodegrade. The ultraviolet light of the sun will eventually break a plastic bag down into smaller and smaller pieces, until it finally disappears. With direct sunlight, this process can be as short as ten to twenty years, not five hundred. The problem is, most plastic bags end up in landfills, where they are soon buried under other trash, and receive little or no sunlight. Suddenly, a thousand years seems a bit more plausible.

Do the environmentalists win this argument? Not so fast. Engineers have developed a way to make plastics out of starches found in corn and potatoes. Not only are these plastic bags more environmentally friendly to produce, but they also can break down in as little as ten days to a month if they are composted correctly—they contain 100 percent cornstarch, with vegetable oil for elasticity. Some companies are switching to these bags, and they soon should be the accepted standard. Whatever the type of plastic—flimsy, durable, or corn-made—it will decompose. It's just a question of when.

Why isn't the whole plane made of the same stuff as the black box?

Planes, believe it or not, are pretty lightweight. They're built with light metals, such as aluminum. The newer ones are built with even lighter composite materials and plastics. This allows them to be fairly sturdy without adding too much weight.

If planes were made of the same stuff as the black box, they just wouldn't get off the ground. But let's backtrack a bit here. The term "*black box*" is a little misleading. What the media refer to as a "black box" is actually two boxes: the Flight Data Recorder (FDR), which records altitude, speed, magnetic heading, and so on; and the Cockpit Voice Recorder (CVR), which records the sounds in the cockpit (presumably ensuring that pilots are on their best behavior all the time). What's more, these boxes are generally bright orange, making them easier to find after a crash. "Black" box either comes from older models that were black or from the charred and/or damaged states of the boxes after a crash.

Whatever the reason behind the name, black boxes are sturdy little things. They carry a bunch of microchips and memory banks encased in protective stainless steel. The protective casing is about a quarter-inch thick, which makes the boxes really heavy. Furthermore, black boxes are not necessarily indestructible—they usually remain intact after a crash partially because they are well placed. They're generally put in the tail of the plane, which often doesn't bear the brunt of a crash.

Even with this extra protection, black boxes sometimes don't survive a plane crash. Still, they typically have been useful, though not so useful that you'd want to build an entire plane with their stainless steel casings.

Is it ever too cold to start a fire?

With the right materials and tools, you can start a fire anytime and anywhere. Fire is simply one result of a chemical reaction between an oxidizer (typically oxygen in the atmosphere) and some sort of fuel (for example, wood or gasoline).

To trigger this reaction, you need to excite the chemicals in the fuel to the point that they will break free and combine with oxygen in the air to form new chemical compounds. In other words, you need to heat the fuel to its ignition point. Once you get the reaction going, the atoms involved will emit a lot of heat, to the point that they glow, producing flames. If the flames from the re-action are hot enough, they will heat more fuel to its ignition point and the fire will spread. The fire will keep itself going until it runs out of fuel.

Cold air does make this process more difficult. Heat energy in the fuel dissipates in the surrounding air, which cools the fuel. The colder the air temperature, the more quickly the fuel will cool. As a result, it takes more energy to heat the fuel to its ignition point in the cold. The two opposing processes go head-to-head: Air is working to cool the fuel while the ignition source is working to heat it up. Whichever process acts more quickly will prevail.

Different types of fuel produce different degrees of heat, which determine their ability to outpace the cooling effect of cold air. For example, propane gas burns much hotter than an equal amount of wood. Heat also increases depending on how much fuel is burning—a burning log cabin is a lot hotter than a single burning log.

Say you are in Antarctica, and it is -129 degrees Fahrenheit (the lowest natural temperature ever recorded on the planet)—you would not be able to get a campfire going by rubbing two sticks together. The cold air would cool the sticks faster than you could heat them to the ignition temperature of wood (four hundred to five hundred degrees Fahrenheit). However, even at that low temperature, you could heat a tank of propane with an electric burner and get a fire going with a spark. You could then use that flame to start a massive wood fire—the collective heat of all the burning wood would outpace the cooling effect of the cold air.

As long as you take along some oxygen, you can even get a fire going in space, where there's no atmosphere to warm things up. Spacecraft use rockets that combine fuel and a compound that contains oxygen to trigger a combustion reaction; the fuel burns, releasing hot gases that push the rocket forward. So with the right equipment, you could theoretically get a campfire going on the moon. You might even be able to roast some marshmallows.

Why are objects in my car's side-view mirror closer than they appear?

Because that's the way lawyers see things. Actually, the printed warning on the glass of your car's right-side mirror has more to do with optics than attorneys, though avoiding liability can never be ruled out where an automaker is concerned.

By noting that "objects in mirror are closer than they appear," designers are telling you the convex shape of the mirror has altered your perception of the distance between you and what's in the mirror. And since what's in the mirror might be a truck, and those of us who are about to change lanes might be fooled into thinking that said truck is farther away than it actually is, a warning is a fine idea.

What's happening is the image of the truck—or car or motorcycle—has been compressed so that the field of vision covered by the mirror can be expanded. The intent is to provide a more complete picture of what's in your over-the-right-shoulder blind spot. This is accomplished by making the right-side mirror slightly spherical in shape.

The advantage is that a wide-angle view takes in more than a flat mirror would. (Try looking into the back of a spoon for a demonstration.) The disadvantage is that when it comes to viewable objects, the mind associates smaller with farther away. Hence, the warning. Incidentally, your car's left outside mirror is planar (flat) because it's closer to your eyes and merely moving your head a bit can expand its field of view. As in your car's planar inside rearview mirror, objects reflected appear near-natural in size.

Car designers have recently introduced aspheric side-view mirrors in which only a portion of the surface has pronounced convexity. Ford is pushing its Blind Spot Mirror, sort of a highly developed version of those little stick-on fish-eye mirrors available at auto parts stores. A convex spotter blended into the top outer corner of the mirror provides a field of view optimized for each vehicle and is usable on left- and right-side mirrors. Already available on some higher-priced cars is a blind-spot warning system that electronically senses unseen objects to the side and alerts the driver with a beeping tone, a flashing light, or both.

Some car companies are even testing video camera systems that would do away with rearview mirrors altogether. Wonder what a lawyer who's had a cable TV signal go down will say about that?

Are electric toothbrushes better than regular toothbrushes?

The evidence says yes, but only if you choose the right type. In April 2005, an independent health care watchdog group, the Cochrane Collaboration, published an analysis of data from forty-two separate brushing studies that had more than thirty-eight hundred participants. Based on the research, the group concluded that electric toothbrushes with circular brushes that rotate in opposite directions are more effective than manual toothbrushes and other electric models. (You can die in peace now, having finally learned the answer to this haunting question.)

So what kind of electric toothbrush should you buy in order

to keep your chompers sparkling? According to the analysis, the aforementioned type with rotating bristles reduced an average of 11 percent more plaque than a regular toothbrush over the course of one to three months. After more than three months of use, rotating-head brushes reduced the signs of gingivitis (gum inflammation) by an average of 17 percent. The analysis didn't find conclusive evidence that other electric models, like the ones with brushes that move from side to side or ionic brushes, are more effective than manual brushes.

But the cheap bastards among us can take heart: Some experts maintain that a good old manual toothbrush is all you really need. In reaction to the Cochrane Collaboration study, the director of the American Dental Association's Seal of Acceptance program, Dr. Cliff Whall, downplayed the benefits of an electric toothbrush. According to Whall, as long as you brush correctly twice a day, floss once a day, and visit a dentist regularly, you're not doing yourself any harm by sticking to the manual method. Now that we've cleared this one up, you can start pondering other all-important topics, like why scratching relieves itching or whether you should wait an hour after eating before you jump into the swimming pool.

How does scratch-and-sniff stuff work?

The basic idea is simple: Manufacturers encase tiny drops of scented oil in thin polymer membranes and stick millions of these little bitty capsules to a piece of paper or a sticker. When you scratch the paper, your fingernail breaks open the tiny capsules and the scented oil (along with its smell) spills out. And there you have it—instant banana odor.

The manufacturing process—called microencapsulation—is a little more complicated. To whip up a batch of scratch-and-sniff stuff, manufacturers combine scented oil and a solution of water and a polymer compound in a big vat. The oil and the solution won't mix, just like oil and vinegar in vinaigrette salad dressing won't mix. But when a giant blending machine stirs everything, the oil breaks down into millions of tiny drops that are suspended in the polymer solution. You see the same effect when you shake a bottle of vinaigrette.

Next the manufacturer adds a catalyst chemical to the mix that reacts with the polymer and changes its behavior. While before it was soluble (it dissolved in water), in its new form it becomes insoluble (it doesn't mix with water). In other words, the polymer separates from the water and turns into a solid. As it solidifies, it forms shells around the tiny oil droplets. The manufacturer dries these capsules and mixes them into a slurry, so that they can be applied to scratch-and-sniff strips.

It's yet another example of how science is being put to good use and making the world a better place for one and all.

Can a person remain conscious after being beheaded?

No, and yes. A person can't remain conscious long enough to plan and exact revenge on the executioners, but it seems that a severed head can get in a final thought or two.

There are many fantastic stories of living, angry heads from the heyday of decapitation. Charlotte Corday, who was executed in 1793 for the assassination of French radical leader Jean-Paul Marat, reportedly blushed when the executioner slapped her severed head. The heads of two rivals in the French National Assembly allegedly spent their last seconds biting each other. And legend has it that when the executioner held aloft the heart of just-decapitated Sir Everard Digby, a conspirator in the Gunpowder Plot of 1605, and said, "Here lies the heart of a traitor," Digby's head mouthed, "Thou liest."

These are extremely tall tales, but more recent accounts are fairly credible. The most famous is from a French physician named Dr. Beaurieux, who witnessed the execution of a criminal named Languille in 1905. Beaurieux wrote:

"The face relaxed, the lids half closed on the eyeballs, leaving only the white of the conjunctiva visible. It was then that I called in a strong, sharp voice: 'Languille!' I saw the eyelids slowly lift up, without any spasmodic contractions—I insist advisedly on this peculiarity—but with an even movement, quite distinct and normal, such as happens in everyday life, with people awakened or torn from their thoughts. Next Languille's eyes very definitely fixed themselves on mine and the pupils focused themselves. I was not, then, dealing with the sort of vague dull look without any expression, that can be observed any day in dying people to whom one speaks: I was dealing with undeniably living eyes which were looking at me."

As horrific as this possibility seems, it is biologically feasible. The brain can still function as long as it receives oxygen delivered via blood. While the trauma of the final cut and sudden drop in blood pressure would likely cause fainting, there still would be enough blood available to make consciousness possible. Exactly how much consciousness isn't clear, but the likely cap is about fifteen seconds.

The next logical question is, what might the beheaded be thinking in these final seconds? Here's a possibility: "Ouch!"

Why can't you tickle yourself?

If only bullies chose this question instead of the far more pop-ular, "Why are you hitting yourself?" You can't tickle yourself because your brain couldn't care less about your attempts at tickling—it basically says, "Duh, I know! I told your fingers to do that." When someone else tickles you, however, the contact is unexpected, and the shock contributes to the effect.

When the nerves of your skin register a touch, your brain responds differently depending on whether you're responsi-ble for it. MRI scans show that three parts of the brain—the secondary somatosensory cortex, the anterior cingulated cortex, and the cerebellum—react strongly when the touch comes from an external source. Think of it like this: When you see a scary movie for the first time, you jump when the maniac suddenly appears and kills the high school kids as punishment for having teenage sex. The second time you see the movie, it isn't a surprise, so you don't jump. The same goes for tickling: It's the element of surprise that causes the giddy laughter of the ticklish.

Why do we laugh hysterically when other people tickle us? Scientists believe that it's an instinctual defense mecha-nism—an exaggerated version of the tingle that goes up your spine when an insect is crawling on you. This is your body's way of saying, "You may want to make sure whatever is touching you won't kill you." The laughter is a form of panic due to sensory overload.

If you're in desperate need of tickling but have no friends or family willing to help, you can invest in a tickling robot. People do respond to self-initiated remote-control tickling by a specialized robot that was developed by British scientists in 1998. There's a short delay between the command to tickle and the actual tickle, which is enough to make the contact seem like a surprise to the brain and induce fits of laughter.

Now that the pressing problem of alleviating loneliness through robotic tickling has been addressed, scientists can shift their attention back to finding a cure for cancer.

Does popping a zit make it go away faster?

Yes, but only if you pop the right kind of zit and do it the right way. What's the right kind of zit? Look for a ripe, juicy white one that appears to be on the verge of a major eruption. Any deep, painful cysts or nodules (you know, the ones with their own pulse), should be left alone or injected with a shot of inflammation-reducing cortisone by your friendly neighborhood dermatologist.

Other pimples to avoid include those above nose-level on your face. Why? The veins around your forehead drain directly into the brain, so squeezing down on zits in this area can actually push bacteria into your gray matter—and you'll end up with much bigger problems than a slightly embarrassing blemish.

Now that that's settled, let's move on to the popping. This is kind of Minor Surgery 101, so please pay attention. First of all, the best time to pop a zit is right after a hot, steamy shower, when your skin is clean and supple and zits have risen to the surface. You might be tempted to just go at 'em with your fingers, but that will only create more redness, swelling, and maybe Freddy Krueger-esque scarring.

The right way to pop a zit—at least according to dermatologist Jeffrey Benabio, who has written scientific articles on this kind of thing—is with a needle (a sharp sewing needle is best). The first thing you must do is sterilize it. Next, take the sterilized needle and hold it parallel to your skin. You're going to lance the pimple at the top of one end and slide the needle through to the other end. Don't worry, this won't hurt a bit—the skin covering the pimple is already dead.

Once the needle is through, gently pull up to open the skin. If the pimple is ripe for the picking, the pus will begin to drain out freely. Finally, clean what's left of the zit with an alcohol-soaked cotton ball or a little witch hazel.

As for those who say you should never pick a pimple, they've probably never had one big enough to inspire total mortification. Truth is, self-grooming is a common practice among all primate animals, not just humans. If you pop your pimples appropriately, they can be off your face—and out of your life—in about seventy-two hours.

How many gallons of pee does the human population produce each day?

Pee pee. Wee wee. Tinkle. Piss. Call it what you will, but everyone does it, every day. Considering there are more than 6.5 billion humans on the planet, there's definitely a whole lot of whizzle whirling around.

How much we each piddle each day depends on a lot of factors—including Big Gulp dependencies and unfortunately enlarged prostates—but we don't need to dive too deep into these issues. In healthy folks, the normal amount of urine passed each day is about equal to the amount of fluids taken in. The Mayo Clinic says for the average adult, that comes out to approximately 1.5 liters a day.

Now this isn't exact science, but to approximate how much pee the human population produces each day, all we have to do is multiply 1.5 liters by the world's population. And here's what we get: 10,016,345,882 liters, 10,016,346 kiloliters, or about 2.6 billion U.S. liquid gallons of pee per day.

Just how much tinkle are we talking? Well, if those 2.6 billion gallons of daily pee were gasoline, you could use it to fuel all the cars in Wisconsin for an entire year. Think about that the next time you point Percy at the porcelain.

Why can't we remember much of anything that happened to us before the age of three?

To spare us the horrific memories of constantly soiling our diapers? Laugh if you will, but Sigmund Freud actually offered an explanation along these lines. Freud, ever the ray of sunshine, believed that we repress our earliest memories because they're uniformly traumatic. (What exactly did this guy's parents do to him?)

Of course, subsequent scientists have had little use for Freud—they've forwarded theories of their own. One post-Freud analysis suggested that young children simply lack the internal equipment to form long-term memories; the prefrontal cortex and hippocampus—the memory centers of the brain—aren't yet developed enough.

Later studies have demonstrated that it's not that simple. Small children actually *do* have the ability to form lasting memories. A 1994 study, for example, found that 63 percent of twenty-three-month-olds could recall events that they experienced at eleven months. (These tiny tots couldn't talk, of course, so the experiment was designed to teach them a unique series of actions and then see if they could still perform them a year later.) This study, and others like it, show that we have the ability to form long-term memories before we can even talk.

In fact, talking is where the trouble begins. According to some studies, the memory starts working differently once language kicks in. Scientists in New Zealand conducted an experiment to observe memory formation in children who were just learning to talk. When the subjects were two or three years old, with limited language skills, they were exposed to a memorable stimulus: a machine that appeared to shrink big toys into smaller toys. (Don't get excited—it was just an illusion.)

The researchers followed up with the kids later, when their language abilities had further developed. While many of them remembered the mysterious machine, they could only describe it using words and actions that were in their repertoire at the time of their exposure to it. Even though they had developed a greater vocabulary that would have allowed them to describe the magical machine with more clarity, they didn't use these words.

This suggests that our early memories become inaccessible because of a change in the way we think. Before we can talk, our worlds—and our memories—are based solely on disconnected impressions of images and sound. As we learn to talk, we develop the ability to tell stories and the ability to structure our impressions into cohesive plots that become narrative, or "autobiographical," memories. While we're learning to talk, we have access to our pre-language memories, but we can't translate them to fit with our new way of thinking. Pre-verbal memories aren't reinforced by and connected to the narrative memories, so they gradually disappear into thin air.

The lesson? Don't bother dropping thousands of dollars on the big Disney World trip before your kid can talk and can remember it.

Why do people yawn?

Chances are good that at some point as you peruse this book, you'll break out into a yawn. It's not because you're anything less than riveted, although it might help if you're a little bored. A yawn is simply one of nature's irresistible urges. Why? Ask a doctor who has spent years studying the structure and function of the human body, and after much theorizing, you'll probably get a dressed-up version of "I dunno."

There are many theories for why we yawn—and why we pandiculate, which is when we stretch and yawn at the same time—but they have holes as gaping as the mouth of an unrepentant yawner.

One theory is that we yawn to expel a buildup of carbon dioxide from our lungs—that the yawn is a maneuver that lets the body take in a larger-than-usual amount of oxygen and displace excess carbon dioxide. But if this were the only reason for a yawn, people who already have enough oxygen would never do it. Studies have demonstrated that increasing the amount of oxygen or decreasing the amount of carbon dioxide in a room does not decrease the frequency of yawns.

Even if it's not entirely satisfying, the oxygen/carbon dioxide theory can be tied to another popular explanation for the phenomenon: boredom and fatigue. When we're tired or bored, we begin to take breaths that are shallower, so it would make sense that we'd need the occasional influx of extra oxygen. But why, then, are yawns contagious? Fifty-five percent of people yawn when they see someone else do it; for many, even thinking or reading about yawning will cause it.

Yet another theory contends that the yawn is a vestigial habit that has been passed down to us from our distant ancestors—a relic that shows how humans communicated before developing language. Perhaps we yawned to intimidate by showing our teeth or as another kind of signal to our fellow humans. Yawning could have been a means of synchronizing a group as it passed from person to person. A similar explanation suggests that yawning actually increases alertness, which could help explain why yawning is contagious—the more alert a group, the more effective it would be at fighting or hunting.

There are also theories that yawning helps cool the brain, keeps the lungs from collapsing, and helps equalize ear pressure. Whatever the reason we yawn, it is certainly ingrained in us from an early age: Fetuses begin yawning at eleven weeks. And unless an eleven-week-old fetus is thoughtful enough to be bored, there's got to be something more to yawning than that.

Why do dead bodies float?

You're in your first year at Gangsters College. After doing a boffo job on your presentation on brass knuckles in your Tools of the Trade 101 class, you've settled in for a riveting lecture on cement boots. And you're shocked—this isn't going to be the breeze you expected. Professor Fat Anthony teaches you that if you don't weigh down a body properly before you throw it into a waterway, it can float to the top. Wait a second—that didn't happen to the guys from *The Sopranos* when they threw Big Pussy overboard!

It's crazy, but true: Bodies that are laden with weight that is equivalent to or greater than the body's shouldn't float to the top. However, bodies that aren't weighted may float for a while. Why? For the most part, it comes down to gas—and not the type that gangsters would use to torch a rat's house. We're talking about gases that form from bacteria in the body during decomposition, including methane, hydrogen sulfide, and carbon dioxide.

Bacteria in our bodies love to eat. When we're alive, they eat the food in our systems; when we die and there is no food left in our systems, they eat us. Bacteria break down what they eat and produce gas. This gas has no way of being expelled from a corpse, so it causes the body to bloat and, thus, float (if it happens to be in water). Once you have a floater, it's going to remain on the water's surface until there is enough decomposition of the flesh to allow the gas to escape.

Not all parts of the body inflate at the same rate. The torso, which is home to the most bacteria, becomes more bloated than the arms, legs, and head. This is partly why a body always floats facedown. The arms, legs, and head can only fall forward from a dead body, so the corpse tends to flip, with the less-gas-filled limbs dangling beneath the giant gas ball of the chest and abdomen.

Depending on the situation, the speed of the decomposition process can vary. For instance, cold water slows down decomposition considerably, while warm water speeds up the bacteria feast. It's a gory sight. Not even the fishes want to sleep with such a thing.

Why do we sneeze?

"Ahhh-chooo!"

"Gesundheit! You just had a sternutation."

"A what?"

"Sternutation. That's the medical term for sneeze."

How did you manage to do that? Chances are, a small particle found its way into your nose. It might have been a bit of pollen, dust, bacteria, a virus, a mite, smoke, or another irritant. Once the nerve cells in the mucous membranes of your nose got wind of it, they released chemicals called histamines. These chemicals acted as messengers, speeding straight to the sneeze center in your brain. (Yes, you really do have a special "sneeze center," located in the area of the brain that connects to the spinal cord, known as the medulla oblongata.) The histamines alerted your brain to the presence of the nasal invaders, and the brain set your body's defense system in motion.

First, your vocal cords closed, causing pressure to build up in your chest and lungs. The pressure built and built and built, and just when you couldn't hold it in any longer, your diaphragm contracted, your eyes shut, your vocal cords snapped open, and air came whooshing out of your mouth and nose at nearly one hundred miles per hour, hopefully carrying the offending particles with it.

Incidentally, sometimes a sneeze begins not in the nose, but in the eyes. If you tend to sneeze whenever you're exposed to bright light, you have a condition known as photic sneezing. Physicians aren't sure why light makes some people sneeze. An estimated 30 percent of the population is thought to suffer from photic sneezing. The condition is not considered dangerous: In fact, some doctors say that glancing quickly at a bright light may be a good way to trigger a sneeze on those occasions when you feel one coming on but can't quite manage to release it.

Is it all right to suppress a sneeze? Sneezes do seem to sneak up on us at the worst possible moments—during a concert, a lecture, or when we're just about to say, "I do." If you hold it in, the resulting implosion could potentially damage small bones in your nose and face or even rupture an eardrum.

So don't be embarrassed. Take a tissue and say, "Ahh-choo." Then blow your nose and say, "I do."

Why does water make your skin pruned?

When you spend enough time in water, the skin on your feet and hands gets wrinkled, or "pruned." No, it doesn't age you: If it did, there would be a lot of filthy people around, clinging to their youth. After you get out of water, your skin eventually returns to normal. The reason for this has to do with how our skin is composed.

Skin is made up of three layers. The deepest layer is subcutaneous tissue that includes fat, nerves, and connective tissue. The second layer is dermis, where you can find your sweat glands, hair roots, nerves, and blood vessels. The top layer (the wrinkle-maker) is the epidermis. The surface—or top outer layer—of the epidermis is made of dead keratin cells. Keratin, which is also part of fingernails, is there to protect the rest of the skin. Your hands and feet have the thickest layer of keratin; since you use these appendages all the time, they need to have an extra protective coating. We couldn't do much if the skin on our hands and feet was as thin as that on our eyelids.

So what happens when you go for a swim or soak in the tub? The keratin absorbs a lot of water. In order to make room for it all, wrinkles form and the skin plumps. Why wrinkling as opposed to just plain ol' swelling? Because the top layer is connected to the other layers of skin, but in an uneven manner. The bottom layers of skin are more water-proof than the top layer, so the water has to sit there for a while before it can be absorbed. Water that is not ab-sorbed by the skin will evaporate, which returns your skin to normal.

The rest of your skin doesn't have such a thick layer of kera-tin, so it isn't as wrinkle-prone. The rest of your body also has hair, at the base of which are glands that secrete an oil called *sebum*. The sebum coats the hair follicles. We all know how oil and water react—the skin does not absorb as much water on your non-hand and non-feet skin. Therefore, the rest of you stays wrinkle-free.

Why does a person get goose bumps?

We've all had the feeling: You get cold or are overwhelmed with a sense of awe, perhaps while watching a fireworks display or a soap opera, and little bumps suddenly appear on your arms, legs, or neck. These are goose bumps, nature's way of saying, "Hey, I'm cold," "I'm f-f-frightened," or "Wow, I can't believe that just happened."

Goose bumps are one of mankind's evolutionary leftovers—bodily structures and functions that were useful at one time in distant prehistory but are now basically pointless, like the appendix. Back when people had more hair, according to one common theory, goose bumps would raise that hair up and trap warm air against the skin to help warm it. Nowadays, only Robin Williams would benefit from this phenomenon.

Goose bumps are an involuntary reflex set off by the sympathetic nervous system. The nerves of the skin cause the little muscles that surround the hair follicles to contract—and you break out in tiny bumps. They're called "goose bumps" because of the way a bird's skin looks when plucked. The term is actually a relatively recent addition to the vernacular—it didn't land a spot in the dictionary until 1933—but "gooseflesh" was used to describe this skin-crawling sensation as far back as the early eighteen hundreds.

And humans aren't the only animals who get it. If you've ever seen a startled cat, you'll notice that the cat's tail appears to become much larger and "poofs" out. This is caused by a reflex that's much like our own—but the cat actually has enough hair for goose bumps to be useful. Many animals appear larger when their hairs stand at attention, which helps to intimidate predators and rivals. And since we also get goose bumps when we feel intense fear or some other extreme emotion, it might have been used to the same effect by our evolutionary ancestors.

Now, though, they are used to help describe a feeling—as in, "It gave me goose bumps"—and not least of all, to name successful scary-book franchises.

What would you look like if they dug you up after you'd been buried for ten years?

The simple answer is: a skeleton. But we know what you really want—you're dying to know all of the gory details about bulging eyeballs and rancid smells. Well, hold on to your barf bags, because you're about to learn the finer points of putrefaction.

First, though, let's clear up a common misconception: Worms don't dine on corpses. Unless a person is buried without a coffin, the main cause of decay is bacteria that are inside the body. These microorganisms exist when a person is living, but the immune system keeps them in check. However, once a person dies and the immune system shuts down, it's open season on the body.

Here's what will happen when you shuffle off this mortal coil. About a week after your death, bacteria will be raging inside your body, and your red blood vessels will begin to rupture, releasing hemoglobin into you. Hemoglobin is the iron-rich element that gives your blood a red appearance. Once the hemoglobin is dispersed, your skin will have the same reddish hue. Eventually, the hemoglobin will break down, turning your skin various shades of green before it becomes dark purple.

A few weeks later, your body will start to ferment, much like alcohol. Fermentation occurs when the bacteria in your body start to break your tissue down into simpler chemical compounds, resulting in the production of gases such as carbon dioxide and methane. Naturally, your body will begin to bloat and take on a puffy appearance. Because the majority of bacteria are in the intestines, most of the swelling will take place in the abdomen. This is also when many of the rank smells associated with death will begin to emerge.

In the final stage of putrefaction, the tissues of your body will completely break down. Your organs will become all sorts of nasty colors and will eventually start to liquefy. Because lean tissue decomposes faster than muscular tissue, the eyes will go first, quickly followed by the stomach and intestines. Once all of the tissue has been destroyed, the skeleton will be all that remains of you. Normally, this process takes about ten years.

Of course, if you're embalmed, it'll be a completely different story. Fluids and gases will be drained from your body, and a disinfecting fluid will be introduced into it. The putrefaction process is much slower; in fact, sometimes the body can be preserved indefinitely.

Just look at Vladimir Lenin. That sucker is holding up pretty well, considering he died in 1924.

What happens to your body when it is struck by lightning?

Alas, of all the firsthand accounts of lightning strikes, nobody has reported gaining any new or exciting superpowers. Mostly, it's just a totally unpleasant experience, even if you are lucky enough to survive it without any major long-term effects. But here's the good news: The chance of being struck by lightning is only one in five thousand, according to the National Weather Service.

Lightning has several ways to get to your tender flesh. It can strike you directly; it can strike an object, such as a tree or another person, then leave it to pay you a visit; it can get you while you're touching something it's striking, like a car door; or it can travel along the ground and take a detour by rising up through your feet. What happens next varies from person to person and often depends upon the intensity of the strike.

Usually, the electrical current travels only over the surface of the skin, a phenomenon called a flashover. This can burn your clothes or, in some cases, shred them off completely and blow your shoes off as well, leaving you in pain *and* naked. Additionally, the metal you are wearing—zippers, belt buckles, jewelry—will become extremely hot, often causing serious burns.

The most immediately dangerous consequence of a lightning strike is cardiac and/or respiratory arrest, which causes most lightning-related fatalities. A strike can also cause seizures, deafness, confusion, amnesia, blindness, dizziness, ruptured eardrums, paralysis, and comas, among other things. Depending on the severity of the strike, some symptoms—such as blindness, deafness, and even paralysis—may disappear quickly.

Contrary to urban legend, lightning does not reduce people to a pile of ash with a hat on top. Additionally, many people believe that lightning-strike victims remain "charged" and are a danger to others after being hit. This is not the case, and this idea too often leads bystanders to delay assistance that could have saved a life.

The most prevalent long-term effects from being struck by lightning are neurological. People can have trouble with short-term memory, distractibility, learning new information and accessing old information, irritability, and multitasking. Multitasking impairment can be especially frustrating.

Many times, tasks that had been easy before the strike suddenly take much longer because the person must focus on every component individually. Damage to the frontal lobe of the brain can cause personality changes, and some victims develop sleeping disorders, cataracts, and chronic pain due to nerve injury.

Why does your stomach growl when you're hungry?

Did you skip breakfast? No wonder your tummy is growling mad. Those gurgly sounds are technically called borborygmi (pronounced BOR-boh-RIG-mee). That's a word the ancient Greeks came up with, and it does a great job of expressing what a stomach growl might actually sound like in spoken form. Go ahead, say it out loud. In case you also skipped English class, it's what academics refer to as onomatopoeia.

But back to your grumbling gut. What's the source of all that clamor? When you haven't eaten for a while, your stomach produces hormones that stimulate local nerves to send a message to the hypothalamus part of your brain. Basically, this is like the hunger light in your head switching from red to green.

In turn, your brain sends a signal back down to the stomach

that says, "Okay, then, get ready to eat!" The result? Muscular activity, and a flow of acids and other digestive fluids in your stomach and intestines. And that's exactly what you're hearing amid all your embarrassment. Your stomach is getting juiced up to chow down.

Amazingly, this connection between your brain and digestive system is so automatic that sometimes the mere thought of food is enough to stir up a snarl. (That's why you shouldn't watch the Food Network at night.) But truth be told, stomach growling can happen at any time, whether you're hungry or not.

Often, the sound is simply a sign of your digestive system at work: muscle contractions and digestive juices churning your last meal into a gooey mix and moving it down the intestinal path. It just so happens that the rumble of your tummy track gets a lot louder when there's not much in there to act as a muffler. So don't ever go to study hall—or a silent movie—without eating first.

Can a person spontaneously combust?

A photo documents the gruesome death of Helen Conway. Visible in the black-and-white image taken in 1964 in Delaware County, Pennsylvania, is an oily smear that was her torso and, behind, an ashen specter of the upholstered bedroom chair she occupied. The picture's most haunting feature might be her legs, thin and ghostly pale, clearly intact and seemingly unscathed by whatever it was that consumed the rest of her.

What consumed her, say proponents of a theory that people can catch fire without an external source of ignition, was spontaneous human combustion. It's a classic case, believers assert: Conway was immolated by an intense, precisely localized source of heat that damaged little else in the room. Adding to the mystery, the investigating fire marshal said that it took just twenty-one minutes for her to burn away and that he could not identify an outside accelerant.

If Conway's body ignited from within and burned so quickly she had no time to rise and seek help, hers wouldn't be the first or last death to fit the pattern of spontaneous human combustion.

The phenomenon was documented as early as 1763 by Frenchman Jonas Dupont in his collection of accounts, published as *De Incendis Corporis Humani Spontaneis.* Charles Dickens's 1852 novel *Bleak House* sensationalized the issue with the spontaneous-combustion death of a character named Krook. That humans have been reduced to ashes with little damage to their surroundings is not the stuff of fiction, however. Many documented cases exist. The question is, did these people combust spontaneously?

Theories advancing the concept abound. Early hypotheses

held that victims, such as Dickens's Krook, were likely alcoholics so besotted that their very flesh became flammable. Later conjecture blamed the influence of geomagnetism. A 1996 book by John Heymer, *The Entrancing Flame,* maintained emotional distress could lead to explosions of defective mitochondria. These outbursts cause cellular releases of hydrogen and oxygen and trigger crematory reactions in the body. That same year, Larry E. Arnold—publicity material calls him a parascientist—published *Ablaze! The Mysterious Fires of Spontaneous Human Combustion.* Arnold claimed sufferers were struck by a subatomic particle he had discovered and named the "pyrotron."

Perhaps somewhat more credible reasoning came out of Brooklyn, New York, where the eponymous founder of Robin Beach Engineers Associated (described as a scientific detective agency) linked the theory of spontaneous human combustion with proven instances of individuals whose biology caused them to retain intense concentrations of static electricity.

Skeptics are legion. They suspect that accounts are often embellished or important facts are ignored. That the unfortunate Helen Conway was overweight and a heavy smoker, for instance, likely played a key role in her demise.

Indeed, Conway's case is considered by some to be evidence of the wick effect, which might be today's most forensically respected explanation for spontaneous human combustion. It holds that an external source, such as a dropped cigarette, ignites bedding, clothing, or furnishings. This material acts like an absorbing wick, while the body's fat takes on the fueling role of candle wax. The burning fat liquefies, saturating the bedding, clothing, or furnishings, and keeps the heat localized.

The result is a long, slow immolation that burns away fatty tissues, organs, and associated bone, leaving leaner areas, such as legs, untouched. Experiments on pig carcasses show it can take five or more hours, with the body's water boiling off ahead of the spreading fire.

Under the wick theory, victims are likely to already be unconscious when the fire starts. They're in closed spaces with little moving air, so the flames are allowed to smolder, doing their work without disrupting the surroundings or alerting passersby.

Nevertheless, even the wick effect theory, like all other explanations of spontaneous human combustion, has scientific weaknesses. The fact remains, according to the mainstream science community, that evidence of spontaneous human combustion is entirely circumstantial, and that not a single proven eye-witness account exists to substantiate anyone's claims of "Poof—the body just went up in flames!"

Why does helium make your voice squeaky?

Everyone loves balloons, especially the ones filled with helium. What's great about helium-filled balloons is not just the cheery way they float along, but also the way the gas inside alters your vocal cords. By inhaling small amounts of helium, a person can change his or her voice from its regular timbre to a squeaky, cartoonlike sound. But how does it work?

The simplest explanation is that since helium is six times less dense than air—the same reason a helium balloon floats—your vocal cords behave slightly differently when they're surrounded by helium. Additionally, the speed of sound is nearly three times faster in helium than in regular air, and that fact lends quite a bit of squeak to your voice as well.

The opposite reaction can be achieved using a chemical known as sulfur hexafluoride, though it's nowhere near as common as helium and much more expensive. Whereas helium is readily available in grocery and party stores, sulfur hexafluoride is generally used in electrical power equipment, meaning that one would have to order somewhat large quantities of it from a specially licensed provider. If you do manage to get hold of some, the results are plenty entertaining: Sulfur hexafluoride drops your voice incredibly low, much like that of a disc jockey or a super villain.

It's important to note, however, that inhaling helium (or other similar gases) is dangerous. There's a high risk of suffocating, because a person's lungs aren't designed to handle large quantities of helium. What's more, the canisters used to fill balloons contain more than just helium—there are other substances in there that help properly inflate a balloon that can be harmful to your body if they're inhaled. So while a helium-laced voice sounds funny, it actually shouldn't be taken as a joke.

Why do your ears pop in an airplane?

Frequent air travelers know they can rely on a few things during each flight. One, a minibag of pretzels and a plastic cup of warm Sprite will be the poor excuse for an in-flight meal. Two, the in-flight entertainment will be a Jim Carrey movie, most likely *Ace Ventura II*. And three, twenty minutes before landing, the infant who has been sleeping peacefully in the row behind you will wake up and begin shrieking nonstop until you land. Is *Ace Ventura II* that bad? Well, yes, it is. But that's probably not why the child is shrieking. More likely, it's because of excruciating ear pain.

Right behind your eardrum is something called the middle ear, a little air-filled space that helps in the transmission of acoustic waves. The air pressure in the middle ear is imperative to your tympanic health—too much pressure and the eardrum could burst; too little and it could collapse.

Usually, that air pressure is pretty stable because of how air pressure works in the environment. As we remember from physics class (okay, as we nerds remember from physics class), air pressure changes with altitude—the higher you get, the lower the pressure.

Whenever air pressure changes, the air pressure in the middle ear also must change to reach equilibrium with the external air pressure. When you're rising—such as during takeoff—the pressure in your middle ear is greater than the pressure outside, and air needs to escape. The escape hatch is known as the Eustachian tube, which connects the middle ear to the throat. During takeoff, this pressure calibration is pretty easy for your ear to achieve on its own. (Think of how easy it is for an inflated balloon to release air.)

However, landing is another story. As a plane descends, the pressure outside becomes greater than the pressure in your middle ear. Left untended, that pressure difference can create a vacuum that makes for a pretty painful earache. Fortunately, we grown-ups know how to force open the Eustachian tube, allowing air to rush into the middle ear and equal the pressure. This can be accomplished by swallowing, chewing gum, blowing your nose, or yawning. The popping you hear during takeoff and landing is the sound of the air rushing in or out of your middle ear.

This also explains why babies tend to shriek on airplanes. Because infants are not capable of willfully forcing open their Eustachian tubes, they simply suffer as the air pressure outside begins to change. One solution is to give the baby a pacifier or bottle, as the sucking motion can help unblock the Eustachian tube and relieve some of that middle-ear pain. If this doesn't work, try turning off the in-flight entertainment.

Why do you stop noticing a smell after a while?

We should all thank our lucky stars for this phenomenon—it makes public transportation a lot more bearable, and it surely has saved countless marriages. And back in our hunter-gatherer days, it made the sense of smell a much more effective survival tool.

To understand why, we need to review the fundamentals of smelling. When you smell something, you're detecting floating molecules that were cast off from all the stuff around you. Inside your nose, you have millions of olfactory sensory neurons, each of which has eight to twenty hair-like cilia that extend into a layer of mucus. These cilia have receptors that detect molecules floating into the nose. Different receptors are sensitive to different types of molecules; for example, when a grass molecule makes its way into your nose, you don't detect it until it bumps into one of the receptors that is sensitive to that particular type of molecule. This neuron then sends a signal to a part of the brain called the olfactory bulb, which is devoted to making sense of odors.

Based on the type of receptors that have been activated, your brain tells you what kinds of molecules are wafting into your nose. You perceive this information as a unique smell. If many sensory neurons fire in response to the same type of molecule, you experience a more intense smell.

This is a handy tool in the wild because it helps you find good food, avoid bad food, and sense predators. And it works a lot better if you're able to filter out ongoing odors that are particularly pungent. For example, you would have a much harder time sniffing out delicious but distant bananas in the jungle if you were preoccupied by the pervasive smell of fetid mud. And since we animals are a naturally stinky bunch, our own body odors would drown out all kinds of useful smells if we didn't possess a means to ignore them.

It's not clear exactly how this happens, but biologists believe that it occurs at both the receptor level and inside the brain. In other words, sensory neurons in the nose reduce their sensitivities to particular types of molecules, while the brain stops paying attention to whatever indications it receives regarding those odors. This filtering process is most likely the driving force behind long-term desensitization to a smell.

It's why a factory worker might stop noticing a strong chemical smell. Or why a wife doesn't run for cover when her husband takes off his shoes after a long day.

Does the body really have its own clock?

It's no crock of New Age malarkey—the body does have an internal clock. Indeed, you don't always eat because your body has run out of fuel, and you don't necessarily sleep because you've used up your energy. A great deal of your behavior is governed by a daily cycle within your body. It's kind of creepy when you think about it—who is in control here? But before you mount a coup against your hypothalamus, learn more about your internal clock.

Your body unconsciously regulates a daily series of processes to help you perform simple functions, such as awakening and remembering to eat. In the early morning hours, for example, your body stops producing melatonin, a neurotransmitter that makes you sleepy, and raises your blood pressure and body temperature in preparation for waking up.

These cycles are not exclusive to humans. All living organisms, including plants, have biological schedules. In some animals, signals from the body can mean the difference between life and death. A nocturnal rodent, for instance, uses the cover of darkness to avoid predators while foraging for food. If that rodent were to suddenly venture out during daylight, it would be at risk from all sorts of creatures.

The fancy name for this internal clock is "circadian rhythm."

The term stems from the Latin *circa* ("about") and *dia* ("day") and refers to the twenty-four-hour period during which these processes take place.

As with any good rhythm, it's a poor idea to suddenly change the tempo. Disrupting your circadian rhythm can trigger health problems, including impaired cognitive function, obesity, and even diabetes. Many things can throw your circadian rhythm out of whack, from jet lag to clinical depression. Up all night cramming for a final? If you feel lousy the next day, it's not just because you're tired—you've derailed your body's natural cycle.

So while the body does run its own maintenance schedule without your knowledge or consent, it still needs your cooperation to get the job done. That means going to bed at a decent hour, setting the alarm, and eating a proper breakfast. Our circadian rhythm is more of a partnership than a dictatorship.

Why do we sweat?

Human sweat glands are like a built-in sprinkler system. Sweat enables us to cool off when the exterior temperature rises (due to changes in the weather) or when our interior temperatures rise (due to exercise, anxiety, or illness). Sweat is one of the mechanisms that our bodies use to keep us at a steady—and healthy—98.6 degrees Fahrenheit.

Here are the basics: Humans have about 2.6 million sweat glands, but not all of these glands produce the same kind of sweat. Sweat has two distinct sources: eccrine and apocrine glands. Eccrine glands exist all over the body and are active from birth. They constantly release a salty, nearly odorless fluid onto the skin, though you probably only notice this sweat when it's really hot or you've been working out really hard. Apocrine glands, on the other hand, are concentrated in the armpits, on the soles of the feet, in the palms of the hands, and in the groin. They become active during puberty. Yes, puberty and perspiration go hand in hand.

Apocrine glands don't secrete liquid directly onto the skin. Instead, each gland empties into a hair follicle. When a person is under emotional or physical stress, the tiny muscle around the hair follicle contracts, pushing the liquid onto the skin, where it becomes sweat. Apocrine glands carry lipids and proteins, as well as water and salt. When these substances mix with the sebaceous oils in the hair follicles and then meet the bacteria on the skin, well, that's when you begin to hold your nose.

But before you start thinking of eccrine as "good" sweat and apocrine as "bad," consider this: Apocrine sweat has been found to contain androsterone pheromones, those mysterious musky odors that are responsible for sexual arousal. So sweat can be sexy, too. Just don't take this as an excuse to wear unwashed gym socks on a date—a few pheromones go a long way.

To banish body odor, a little dab of deodorant should do. Deodorants are based on mildly acidic compounds that dry the skin before the odor starts. Antiperspirants, another popular option, actually block sweat with aluminum salts. Some people think that these salts may be unhealthy, but so far, clinical evidence has failed to connect them to any disease.

If you feel that you sweat too much, or too little, see your doctor. Excessive sweating, officially known as *hyperhidrosis,* and lack of sweat, called *anhidrosis,* are genuine medical conditions with serious complications. Fortunately, both are treatable. For most of us, however, dealing with sweat is fairly simple: Take a shower and wear loose and absorbent clothing. For goodness sake, don't sweat about sweat!

Can you walk on hot coals without burning your feet?

There was a time when the feat (pun most certainly intended) of walking on hot coals was the domain of mystical yogis who dedicated their lives to pushing the physical limits of the body by using the awesome power of the mind.

Then along came reality television. Now on any given night, we can tune in to some pudgy actuary from Des Moines waltzing across a bed of glowing embers for the nation's amusement, seemingly unharmed. So what's the deal? Is walking on hot coals dangerous, or even difficult?

At the risk of prompting legions of idiots to inflict third-degree burns on themselves, the answer is no. Walking on hot coals is not as impressive as it seems—but please, please, read on before you try something stupid.

The secret to walking on hot coals has nothing to do with mental might and everything to do with the physical properties of what's involved. It comes down to how fast heat can move from one object to another. Some materials, like metal, conduct heat very well—they're good at transmitting thermal energy to whatever they touch. Think of your frying pan: You heat it up, slap a juicy steak down on it, and witness an instant sizzle—the metal easily passes its heat to an object of lower temperature. On the other hand, consider the bed of hot coals that's used for fire walks. It started out as chunks of wood—and wood is a terrible conductor of heat.

But don't go for a romp over hot coals just yet. It's also important that the hot coals are not, you know, on fire. If you've seen a fire-walking demonstration on TV, you may have noticed that there were no jumping flames, just smoldering embers— the coals probably had been burning for hours and had built up a layer of ash. And ash is another poor conductor of heat—sometimes it's used as insulation for this very reason.

But all the ash in the world can't help you unless you keep one final thing in mind. Think about it: What sort of gait do you see when a person is traversing a bed of hot coals? A stroll? An amble? A saunter? No, no, and no—it's all about making a mad dash. As a result, the amount of time that any one foot is in contact with a coal might be less than a second. And the exposure is not continuous, as each foot gets a millisecond break from the heat with each step.

So if you take a poor heat conductor like wood, cover it with a layer of insulation, and have intermittent exposure to the heat, the likelihood of sustaining serious burns is low. Of course, we don't advise that you try this stunt at your next backyard get-together. What if you fall? Or even slip or stumble? You'll have a lot of explaining to do at your local ER.

Why doesn't your stomach digest itself?

Your stomach is the fourth stop on the amazing journey along the digestive tract (after the mouth, pharynx, and esophagus), but it's the point where the system gets down to serious business. Cells in the stomach produce two to three quarts of hydrochloric acid and digestive enzymes daily, and muscles work to churn everything together in order to create a soupy goo. This digestive gastric juice is potent stuff— it's strong enough to break down wood and metal, let alone food. And the only thing that keeps it from eating through your body is a thin layer of equally powerful mucus.

Acids are dangerous because when they are dissolved in water, they release excess hydrogen ions (hydrogen atoms with a positive charge). These ions react easily and quickly with other material, breaking chemical compounds down into simpler compounds. In the stomach, the hydrogen ions combine with protein compounds in food to form amino acids and simpler polypeptide compounds that the small and large intestines can further digest.

The stomach wall is made up of vulnerable proteins, which means that you would be in big trouble if hydrochloric acid were to reach it. Luckily, the stomach surface is lined with cells that continually secrete mucus that neutralizes the acid; this mucus is loaded with bicarbonate, a powerful chemical base. (A base is essentially the opposite of an acid.)

When bicarbonate is dissolved into water, it results in the release of hydroxide ions, which have a negative charge. Negative hydroxide ions and positive hydrogen ions effectively cancel each other out if they are combined; the chemicals undergo a reaction and form water and other harmless products.

The system works well, but since it depends on constant chemical balance, it isn't entirely foolproof. Sometimes acid will erode part of this mucus, resulting in a painful gastric ulcer in the stomach lining. In extreme cases, the acid will erode a hole all the way through the stomach wall, and the stomach's contents will spill out into the abdominal cavity.

Fortunately, the stomach lining keeps most of us safe, even when we eat like pigs. So next Thanksgiving, remember to count mucus among your blessings.

Why aren't people covered in hair like other primates?

Life would be much more confusing at the zoo, for one thing. Cover us in fur and the line between ape and man gets a lot blurrier.

Scientists have proposed a few explanations for why people are mostly hairless. In the 1970s, the hot theory was that early humans entered a semi-aquatic phase—they spent their days catching fish and eating plant life in shallow waters. Like whales and hippos, humans lost their fur in favor of a layer of fat under the skin, which is a better insulator in the water.

But there are some holes in this theory—the most notable being that hanging out in African lakes and rivers is dangerous, thanks to crocodiles and nasty water-born parasites. If we had spent that much time in the water, critics say, we would have developed better defenses against the parasitic worms and such that still kill people in Africa today.

Speaking of parasites, a more recent theory speculates that we lost our fur as a defense against lice, fleas, and the like. When you're covered in fur, parasites are a major problem—but the insulating benefits of a flea-ridden coat of fur make the problem worth enduring. Until you've invented clothing, that is. As soon as humans could make their own clothes and build their own shelters, fur became a liability.

The problem with this theory is that the timing doesn't work—scientists have traced our fur loss back about 1.7 million years, yet evidence of clothing goes back only about forty thousand years. Still, even if parasite-avoidance wasn't the driving force in our loss of fur, it might help to explain why we think smooth, hairless skin is sexy today. Think back to our caveman days, and you'll see why we might have been hardwired to favor hairlessness. If you had no body hair, a potential mate could clearly see whether you had parasites; if you didn't, it suggested that you were healthy and would produce strong offspring.

But back to how we lost our fur in the first place. The most popular theory nowadays is that the scorching savannah was just too darn hot. As humans gradually developed a new foraging and hunting lifestyle, they began leaving the cool jungles behind for long treks in the hot sun.

And then, just as now, humans cooled off by sweating. As our hirsute ancestors chased their prey across the vast savannahs, they would have soaked their furry coats in no time, making it even more likely that their bodies would overheat. Having no fur enables sweat to cool the skin—and, thus, the body—directly and effectively. In our ancestors' new broiling environment, the benefits of keeping cool in the daytime would have outweighed the problem of poor insulation at night. Meanwhile, our primate cousins never left the cool jungles, so they never made the shift to hairlessness.

Humans retained the hair on their heads, it seems, because these locks help to keep the brain cool—when the sun beats down on our heads, it heats up the outermost layer of hair rather than fry the scalp directly. As for the stuff on our armpits and around our groins, the leading theory is that this hair accentuates pheromones (chemicals that are supposed to entice the opposite sex through the sense of smell). Who knew that B.O. is sexier than perfume and cologne?

Can you really pick up radio stations on your dental work?

If you've watched enough *Gilligan's Island*, you know that it can happen on TV. And maybe you've heard that it happened in real life to Lucille Ball of *I Love Lucy* fame. According to Jim Brochu, author of *Lucy in the Afternoon*, the actress claimed that her dental fillings picked up radio signals as she drove to her home outside Los Angeles in 1942. She also claimed that the signals were later traced to a Japanese spy who was eventually taken into custody by law enforcement authorities, perhaps the FBI.

Who knows if Lucy's tale is true? Nobody's found documented evidence of a Japanese spy nest infiltrating California in 1942, and Lucy's FBI file contains no mention of such an event. (Yes, Lucy had an FBI file. At the urging of her grandfather, she had registered to vote as a Communist in the 1936 elections, so she had some 'splaining to do when she was investigated by the House Select Committee on Un-American Activities in 1953.) Recently, the Discovery Channel television show *MythBusters* devoted a segment to debunking Lucy's claims.

Other people have claimed that they picked up radio signals via the metal in their heads, whether it was dental work or something else. The anecdotal accounts are easy to find but hard to verify. In 1981, however, a doctor in Miami wrote to *The American Journal of Psychiatry* to report that he had treated a patient who suffered from headaches and depression and complained of hearing music and voices. The patient was a veteran who had been wounded by shrapnel to the head during combat twelve years earlier. But after receiving successful treatment for the headaches and depression, the patient claimed that he still heard the mental music.

This led the doctor to sit down with the patient and a radio; the two of them listened to various stations, trying to find one that matched what the patient heard in his head. When they found what seemed to be the offending frequency, the doctor listened to the radio with an earphone while the patient described what he was hearing in his head. Although the patient couldn't hear the voices clearly, he passed the test convincingly; he was even able to tap out the rhythm and hum along with the songs that played. The doctor concluded that the shrapnel in this man's head was receiving radio signals and conducting the sound through bone to his ear.

So apparently, it is possible for the metal in your head to receive radio transmissions—but don't look for the American Dental Association to begin marketing the iTooth portable music player anytime soon.

If you donate your body to science, what do they do with it?

You can rest assured that scientists don't take donated cadavers out for wild Weekend at Bernie's-style partying or prop them in passenger seats just to use the carpool lanes. Typically, donating your body to science means willing it to a medical school, where it will be dissected to teach medical students about anatomy.

Fresh cadavers aren't as critical to medical schools as they once were, thanks to detailed models, computer simulations, and a better ability to preserve corpses. But they're still a much-appreciated learning aid. If you have a rare deformity or disease, your corpse will be especially useful.

Medical schools aren't allowed to buy bodies, rob graves, or go door to door recruiting volunteers, so they rely on potential donors to initiate contact. If you want to donate your body, you'll need to find a medical school in your area that has a body donation program. Your state's anatomical board is typically a good place to start. Once you've found a program, you fill out some legal paperwork and perhaps get a body donor identification card to carry in your wallet. Some schools will cover the cost of transporting your corpse to the school, within a certain distance, as well as cremation costs; others won't pay for transportation.

This is very different from organ donation, which you can arrange in many states by adding a note to your driver's license and sharing your wishes with your family. If you're an organ donor and die under the right circumstances (you're brain-dead but on a respirator), the doctors may extract your heart, kidneys, lungs, liver, pancreas, or small intestine and whisk the pieces to the organ recipients. But if you aren't on a respirator when you die, these internal organs won't be usable.

If you've already donated your organs, most medical schools won't accept what's left of your body. You're also out of luck if you died from a major trauma, had a contagious disease, or underwent major surgery within thirty days of your passing. And if you were obese or emaciated or your body has deteriorated? Again, you're out of luck.

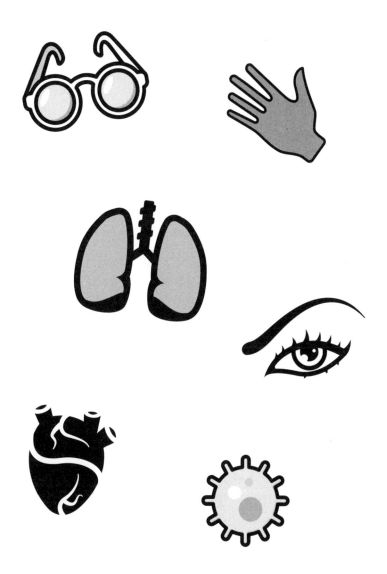

Odd Things About Animals

Can worms cure allergies?

In the dark ages of medicine, the leech–that sloppy, sludgy parasite that looks sort of like a worm on steroids–was used as a tool to fight illness. Of course, we know better now–our high-tech medical remedies are synthesized in sterile labs, not dredged up from muddy lakes. Right?

Not exactly. It turns out that disgusting parasites may still be useful in treating certain conditions that have baffled modern medicine (more useful, in fact, than leeches, which were never really an effective remedy). In 2004, Dr. David Pritchard, an immunologist at Britain's University of Notting-ham, deliberately infected himself with hookworms as part of an experiment concerning allergies. And it worked.

Hookworms are nasty little parasites. An untreated infesta-tion can lead to severe anemia and even death. Hookworms are common in undeveloped tropical countries, where it just so happens that allergies are relatively rare. In the devel-oped world, the problems are just the opposite: Few people need to worry about parasitic worms, but allergies are more and more common, causing at least two million emergency room visits each year.

While conducting research in Papua New Guinea, Pritchard began to wonder if this was more than coincidence. Are worms a key to preventing allergies? To test his hypothesis, he gave a group of villagers pills to kill their worms and asked if, in return, they would give him what he politely termed their "fecal matter" to study. Since Pritchard and his assistants didn't speak the local language, they must have used some interesting gestures to get their point across. But it turned into a win-win situation for both parties—the New Guineans got health care and Pritchard got scientific evidence. By comparing the number of worms eliminated by each patient with the concentration of antibodies in the bloodstream, he could see how worms affected the immune system. Since an allergic reaction is basically the body's attempt to defend itself against a hostile invader, worms must be able to suppress this response in order to survive in the gut.

This ability intrigued Pritchard. People with a lot of allergies have hypersensitive immune systems—perhaps a few worms might lower these from constant states of "red alert" to something more akin to "yellow" or even "green." Back home in Britain, Pritchard infected himself to prove that a small dose of worms was not dangerous to the average healthy Westerner. After that, he was able to obtain funding and thirty willing volunteers for a clinical trial. Within a week, the fifteen subjects who had ingested a mere ten worms each found their allergy symptoms disappearing. The others, who had received a placebo, showed little or no improvement. When word of these results hit online discussion boards, inveterate sneezers began demanding Pritchard's "helminthic therapy" to ease their symptoms.

If you have hay fever but are not keen on ingesting hookworms for relief, wait a few years. Scientists are trying to isolate the specific substance that worms use to disarm allergic reactions and are aiming to produce it in pill form, a much more palatable way to fight allergies than embracing our inner parasites. Hopefully, someday soon, you'll be able to pop one of these pills, get a good night's rest, and be fresh as a pollen-proof daisy by morning.

Can it rain fish?

It sure can. While it's a rare occurrence, there are dozens of fishy rainstorms on record. For example, in July 2006, a downpour of pencil-size fish pelted residents of Manna, India. It seems that Mother Nature is one mad scientist.

Long ago, people attributed these fish-storms to the wrath of God or mysterious oceans in the sky. But today, most scientists agree that waterspouts are the actual culprits, though this hypothesis hasn't been definitively proven.

Waterspouts are essentially weak tornadoes that form over

bodies of water. Some waterspouts occur in the midst of thunderstorms, similar to the tornadoes that appear on land, but most are fair-weather creations—weaker funnels that can occur even on calm days. Sometimes, the theory goes, these funnels suck up water—and any creatures that happen to be swimming around in it—from the surface of an ocean or lake. Air currents can keep the fish aloft in the clouds before dropping them onto unsuspecting people who are up to a hundred miles inland.

It's difficult to say how often this freak show occurs, but in some places, it's believed to be an annual event. Residents of Yoro, Honduras, claim that each year between June and July, a big storm leaves the ground covered with tiny fish. These critters may not actually fall from the sky, though; some zoologists theorize that the Yoro fish actually come from underground streams and are stirred up by the rain.

Besides fish, frogs are probably the most common animals to rain from the sky—a plague of falling frogs even makes an appearance in the Old Testament. In July 2005, a mysterious cloud dropped thousands of live frogs onto a small town in Serbia. (Perhaps a modern-day Moses was in its midst.)

After fish and frogs, it gets even stranger. Some of the more noteworthy bizarre rainstorms: jellyfish in Bath, England, in 1894; clams in a Philadelphia suburb in 1869; and spiders in Argentina's Salta Province in 2007. We'd be willing to bet that the local weathermen didn't forecast these storms.

Why is bird poop white?

Since a typical bird dines mainly on worms, bugs, and as-sorted garbage, it seems logical that it would poop in hues of brown, like the rest of us. What's going on? Do birds take special white-poo supplements just to maximize their car-defacing potential? No. Bird poop is, in fact, brown. The white stuff is urine—and this leads us to a sophisticated discussion about the differences between pee and poop.

You pee to eliminate excess water and waste products that result from cell metabolism. Essentially, the cells in your body do what they do through chemical processes that use the oxygen you breathe and chemical compounds derived from the food you eat. These chemical processes result in new, leftover chemical compounds that your body needs to expel. You eliminate one of the biggies—carbon dioxide—through breathing. You eliminate everything else through sweating and peeing. Poop, on the other hand, is food that your body didn't need in the first place. The body takes what's useful from the food that you eat, breaks it down into sugars and proteins your cells can use, and sends the rest on its way, as poop.

Among the primary metabolic waste products your body must get rid of are nitrogen compounds. These compounds are toxic to cells and would kill you if they remained in your body in their straight form. Your body needs to convert these toxic compounds into something safer en route to their excretion. Humans, other mammals, and amphibians do this by turning nitrogen compounds into a substance called urea, which can be dissolved into water. When converted to urea and added to water, the nitrogen is relatively harmless.

This doesn't work for reptiles and birds, however, because they lay eggs. A bird embryo that's inside a shell doesn't receive a steady supply of water, and since it has no place to keep all that urine anyway, urea won't do the trick. Instead, reptiles and birds convert their waste into uric acid—white crystals that form a pasty solid, which a bird embryo can safely have inside the egg until it's hatched. The uric-acid excretion process means that birds and reptiles don't have to drink much water—they need just a bit of it to excrete the urea.

In birds, poop proper comes out the same hole as the uric acid waste, so most droppings are a mixture of the white uric acid and brown material. And that, fair reader, is why the stuff that's clinging so stubbornly to your car's windshield looks the way it does.

Can chickens really run around after their heads are cut off?

You bet they can! If you thought this was just an expression for being in a frenzied state, think again.

When a chicken is decapitated, its brain is severed from its spine, and any voluntary control over movement ceases. However, a chicken's involuntary movements—say, running around like it just lost its head—are controlled by electrical impulses, which originate at the spinal cord and continue until the muscles themselves run out of energy. As for the length of time a chicken can run around post-beheading, that depends on the chicken.

If your stomach isn't already turning, chew on this less-than-savory story about the famous Mike, a Wyandotte rooster that lived for a whopping eighteen months after its head was detached. On September 10, 1945, Mike's owner, farmer Lloyd Olsen of Fruita, Colorado, quite awkwardly tried to behead the five-and-a-half-month-old feathered fellow in preparation for a meal for his mother-in-law. Alas, Olsen missed the jugular vein (a blood clot prevented Mike from bleeding to death), and left one ear and most of the brain stem intact. That night, poor Mike slept with its severed head tucked beneath its wing.

Mike survived, and attempted to walk, hunt, peck, preen, and gurgle in place of crowing. A milk-and-water mixture and some grains were fed to the bird via an eyedropper, helping it grow to a strapping eight pounds. Olsen and his family molded a successful showbiz career for Mike, which included twenty-five-cent sideshow appearances across the nation and photographs for popular magazines, such as *Time* and *Life*.

Mike earned up to $4,500 a month and was valued at $10,000. But don't cry fowl at these exploitations and indignities—it was the consensus of humane society groups that the rooster didn't suffer. Unfortunately (or fortunately), Mike met his demise in a Phoenix motel in March 1947, when the Olsens were unable to swiftly locate an eyedropper to clear mucus from Mike's throat.

Today, residents of Fruita pay their own wacky tribute to the fallen rooster with a Mike the Headless Chicken Web site (www.miketheheadlesschicken.org), complete with pictures of the headless hero strutting his stuff. They even celebrate Mike's unusual life with an annual festival held the third weekend in May. Considering all of this fuss, the question needs to be asked: Who exactly lost their heads?

When lightning strikes the ocean, why don't all the fish die?

For the same reason we don't all die when lightning strikes the ground: The ocean and the ground both conduct electricity relatively well, but the current from a lightning bolt dissipates quickly as it spreads through the earth or through a large body of water. Only the area surrounding the strike feels the shock.

Thanks to all of its dissolved salt and other impurities, seawater is a good conductor. The charge from a powerful lightning strike could spread out more than a hundred feet, and any fish in the immediate area would probably get zapped—but only if it isn't swimming too deep. This is because electricity likes to flow along the surfaces of conductors rather than through their interiors, so when lightning strikes the ocean, most of its current spreads out over the water's surface. And even if some fish are near the surface, they won't necessarily take the full brunt of the charge. Electricity follows the path of least resistance, and seawater conducts currents much better than fish do—in other words, the electricity would want to flow around the fish rather than through them. Even so, if a fish happens to be swimming too close to the site of a powerful strike, the jolt will be deadly.

Fortunately for fish, lightning strikes the ocean far less frequently than it hits land. One of the conditions that makes thunderstorm formation possible is the rapid heating of low-lying air. But oceans don't reflect nearly as much heat as the ground does, so the atmospheric trends that exist over the ocean aren't particularly conducive to forming thunderstorms.

But don't take this information as clearance to run into the ocean when a storm is brewing. When there's lightning around, you want to be surrounded by insulators (like your house), not dog-paddling in a giant electricity conductor.

Do cows cause global warming?

At first glance, cows embody the ideal for a clean, healthy planet. What could be better than a farmer working the land and raising animals in a lush, green meadow? But what's that sound? And—whoa!—that smell?

In addition to milk and meat, cows give us burps, farts, manure...and, in turn, methane. A cow has a four-chambered stomach. Food is partially digested in the first two chambers and then regurgitated for cud-chewing; the food is then fully digested in the third and fourth chambers. The food ferments as it is digested, and the gases expelled during this process (from either end of the cow) include methane. The "cow patties" the animal produces also contain methane.

Of the greenhouse gases blamed for global warming, only carbon dioxide is considered more harmful than methane—carbon dioxide gets top billing because of the sheer volume of it floating around the atmosphere and trapping the sun's heat. Methane is at least twenty times better at trapping heat than carbon dioxide, and the volume of it that's in the atmosphere has almost doubled in the past two centuries. This, however, can't be blamed solely on cattle—landfills, decomposition in wetlands, coal mining, petroleum drilling, and even rice paddies produce large amounts of methane, as well.

Nevertheless, farm animals are definitely doing their share.

Globally, about 15 or 20 percent of methane comes from livestock, including the world's more than one billion cows. Each of these cows has the potential to fill about three hundred two-liter bottles with gas each day. Once the methane produced by other four-stomached, or ruminant, livestock—including sheep, goats, and buffalo—is factored in, the annual global total is eighty million metric tons. This makes livestock one of the largest human-related sources of methane (as we're the ones feeding and raising them). In the United States alone, livestock account for about 20 percent of the methane released into the air.

Cows aren't going away anytime soon, so resourceful farmers and scientists are looking for ways to minimize the problem. For instance, improved nutrition for cows—feeding them more easily digestible grasses, for example—can help diminish their methane excretion. Scientists are also developing a new strain of grass that may further reduce cows' propensity toward toxic emissions.

In the meantime, the methane trapped in cows' solid waste is easy to harness and use productively. Methane is quite similar to the natural gas used for fuel, so some farms have taken to collecting cow poop, extracting the methane, and burning the gas to create electricity. As an added bonus, the remaining components of the manure can be used as fertilizer (the liquid) and as compost or even bedding (the dry remnants) for the very cattle that produced it.

Do migrating birds ever get lost?

That depends on what you mean by "lost." Migrating birds sometimes stray from their usual course due to powerful winds and harsh weather, so they might end up thousands of miles from their intended destination. But even then, the migrating bird is never truly disoriented. Birds are known to traverse oceans by accident or swap continents without ever planning to do so, but often these misdirected travelers return to their original nesting grounds the following year. Such is the go-with-the-flow lifestyle of our feathered friends: A gust of wind in the wrong direction and they're spending the winter on West Africa's Gold Coast instead of in the Florida Keys. No biggie.

But how do migrating birds always know where they are, even when they're in the wrong place? They use visual cues, along with the mysterious force of geomagnetism, to chart their travel.

During the day, birds can look at the movement of the sun to ascertain their position. Homing pigeons, for example, use the sun as a compass. In an experiment in which homing pigeons were exposed to an artificial, twenty-four-hour light, the birds lost track of the sun's location. Upon their release, they were unable to navigate properly.

Night-flying birds use the stars to orient themselves; like sailors of old, they use the position of the North Star as a guide. This form of visual navigation has also been verified by a scientific experiment. Two sets of birds were kept in a planetarium. One set was shown constellations revolving around the North Star, as per usual; the second set was shown constellations revolving around Betelgeuse (a different star). Upon release, the first set was able to orient itself properly, but the second was not.

There are other, more down-to-earth sights that help birds plot a trip. It is common to see birds following a coastline, the meandering path of a river, or the course of a mountain range.

When visual aids fail, migrating birds have another tool at their disposal: geomagnetism. Many animals in addition to birds—salmon and bats, for example—have small amounts of magnetite inside their heads. This element reacts to changes in the earth's magnetic field and gives the animal an innate sense of direction, much as the fluctuation of fluid in our inner ears gives us an innate sense of balance. Some scientists believe that birds even use their sense of smell to enhance their mental maps.

But how do birds decide where they want to go in the first place? Some scientists believe it comes down to hereditary memory, which would inform a youngster of the route it must take to join the rest of its species at breeding time. There are enough examples of abandoned young making such journeys for this explanation to seem plausible. For example, many species of birds do not bring their offspring along when they migrate; they leave them to find their own way.

Can you blame them? Round trips for some birds can be as long as twenty-five thousand miles. Imagine the little birds continually chirping, "Are we there yet?" As humans know, it would be enough to drive the big birds crazy.

Can fish walk?

It's hard being a big fish in a small pond. When food gets scarce, there's no place to go. Unless, of course, you are a *Clarias batrachus, Anabas testudineus,* or *Periophthalmus modestus.* In that case, you just climb onto dry land and go looking for new digs.

The walking catfish, climbing gourami, and mudskipper, as they are respectively commonly known, are the three main species of ambulatory fish—fish that leave the water voluntarily rather than at the end of a line.

The walking catfish, which can grow to be nearly a foot long, has an omnivorous appetite and a nasty sting, and is the most notorious of the bunch. Native to southeast Asia, it is also found in Sri Lanka, eastern India, and the Philippines. It arrived in the United States in the early 1960s as an aquarium fish that was imported to southern Florida by exotic animal dealers. A few escaped into the wild, and by the 1970s they had spread to freshwater ponds throughout the state.

In reality, "walking" is not quite how this catfish gets around. The Thai call it the "dull-colored wriggling fish," a far more accurate description. On land, the catfish propels itself along with a snake-like motion. It breathes though labyrinth organs, which are located above the gills and absorb oxygen from the air. How far can these creatures walk (or wriggle)? A few yards at the most, according to observers.

The climbing gourami, another freshwater denizen, does the walking catfish one better. Originally from Africa, these fish can now be found in India, Malaysia, southeast Asia, and the Philippines. The gourami uses its gill plates, fins, and tail as primitive legs, and has been reported to climb over small trees on its journeys from pond to pond. Like the walking catfish, the gourami breathes through labyrinth organs and can survive for several hours on land, as long as its skin remains moist.

The third member of this trio, the mudskipper, is the champ among walking fish. It is the most widespread, too, found on the coasts of west Africa, Australia, Madagascar, India, Japan, Indonesia, and the Philippines. Genuinely amphibious, the mudskipper seems just as comfortable on land as in the water. These fish venture ashore for extended periods of time, using their strong pectoral fins and tail for locomotion. They can even flip themselves like acrobats. Oxygen is absorbed directly through the mudskippers' skin by a process known as cutaneous air breathing. In fact, the mudskipper is so well adapted to land, it will drown if it spends too much time submerged in the water without a trip to the surface for a breath of fresh air.

So if you happen to see a walking fish on your next tropical vacation, it's not because you had a few too many mai tais. That fish is probably out taking a stroll when the tide is low. Just like you.

Is spider silk really the strongest material in the world?

No—but it comes remarkably close.

The strongest material in the world has been recognized as the carbon nanotube, which is fifty times stronger than steel. These are microscopic, tubelike molecules of carbon, smaller in diameter than a strand of hair.

Spider silk, on the other hand, is five times as strong as steel and is comparable in strength to Kevlar (the material used to make bullet-proof vests). Spider silk can stretch to 130 percent of its length before breaking and is estimated to withstand up to six hundred thousand pounds of pressure per square inch.

A composite material made up of crystalline molecules, spider silk puts synthetic materials to shame in terms of the range of applications for which it might be used. "Liquid crystal," as it's called by the science community, has achieved perfection after four hundred million years of evolution, and lately it's getting scientists' panties in a bunch.

Spider silk has been studied with envy for some time now, and there have been various attempts at domesticating spiders with the intention of harvesting the material. None of these have panned out, mostly because spiders cannot stand to live in close quarters with other spiders. They need room to live, and they will kill any other spider that cramps their space. Because of this, spider farming was ruled out of the question early in the silk-production game.

But a Canadian company called Nexia Biotechnologies has lately stumbled upon an answer. Through bioengineering, the company's scientists have been able to produce mass quantities of orb spider silk in the mammary glands of goats. The scientists take cells from one goat, mix them with specific spider genes in a culture dish, and then transfer the cells into unfertilized eggs. (The process is similar to that which produced Dolly, the world's first cloned sheep.) The resulting male goats are then put out to stud. Females born from this breeding produce spider-silk proteins in their mammary glands. These proteins are extracted from the milk and mechanically spun into fiber.

There has been success in fostering a spider's silk proteins in other mammals—and in some plants, to a limited extent—but goats have been particularly adept at producing the proteins. The speed at which female goats mature, and the ease of caring for the animals, likely played a part in their selection for this particular task.

Nexia intends to put the material to a variety of uses, including fishing line, sutures, tennis racket strings, body armor, and even replacement tendons. Spider silk appears to be the perfect material for each of these tasks, based on strength, flexibility, and biodegradability.

It's not perfect yet—natural silk contains seven different proteins, while the silk produced in the goats' mammary glands contains just one—but Nexia's scientists are working on the problem. While they're at it, they might want to give some thought to the company's marketing plan. Spiderweb tennis rackets will be all the rage until the players get out on the court and find they can't even complete a single volley. Soon to follow will be a wave of rackets returned to the manufacturers with tennis balls stuck in the middle of the webbing.

How does an eel generate electricity?

At the risk of being particular to the extreme, eels don't generate electricity. The animal commonly known as the electric eel (*Electrophorus electricus*) is actually a member of the Ostariophysi superorder of fish. These critters look like eels, but they reside primarily in fresh waters. True eels, on the other hand, spend most of their lives in the ocean.

Now that we're clear—how does *Electrophorus electricus* create its charge? Bioelectricity!

The electric eel is one of a small group of fish that has an electric organ, located in its tail portion, which is most of its body. Within the organ are about six thousand flat cells, called electrocytes, that are arranged in columns, like a stack of plates but with a small amount of fluid between them. These six thousand-some electrocytes are very excitable, electrically speaking, and are not unlike a large group of household batteries serially connected to each other with the positive pole at the head and the negative pole at the tail. They wait for a signal to emit a charge into the surrounding water and when the electric eel's brain sends the message to "fire!" the electrocytes discharge almost simultaneously in an extraordinarily high-speed domino-like reaction.

In two to three milliseconds, a brief but powerful electric current courses along the eel's body. In that moment, the eel can generate as many as six hundred volts—a jolt five times more powerful than that from a standard wall socket and great enough to seriously injure a human—via rapid-fire, pulsing electric organ discharges (EODs). Mostly electrically inert when not in motion, the electric eel emits twenty to thirty EODs per second when it starts moving and can hit fifty EODs at peak periods.

Why do electric eels emit electricity? Food, protection, and navigation. When it comes to dining, the amount of electricity the electric eel delivers usually won't kill a large fish, but it will stun it for long enough to allow the electric eel to eat in peace, without having to deal with all the thrashing. This electricity also serves to stun would-be predators. Lastly, the electric eel finds its way about its habitat, mainly the Amazon, by emitting EODs that detect objects around it through a process called electrolocation.

Scientists are studying the electric eel and other Ostariophysians, such as the knife fish, to learn more about their production of electricity and the uses we might have for it. So far, however, this has amounted to little more than sideshow attractions. In an aquarium in Japan, for instance, one electric eel helped light a Christmas tree in December 2007. It might seem small—one fish, one string of lights—but the implications of a living, breathing source of energy are pretty fantastic.

What do weasels have to do with coffee?

So you've heard the stories and you swear that you'd never drink coffee made from beans that have passed through a weasel's digestive tract. Luckily, you'll never get the opportunity. Those ultra-gourmet beans that sell for anywhere from one to six hundred dollars a pound didn't come out of a weasel. The animal in question is actually the Asian palm civet. Does that make you feel better?

No? Perhaps you're like columnist Dave Barry, who dismisses the stuff as "poopacino" and thinks the whole craze for exotic brews is nothing but a tempest in a coffee cup. But kopi luwak, as it is known in Indonesia, got a big thumbs-up from Oprah Winfrey in 2003 when representatives of the Coffee Critic, a Ukiah, California, coffee shop, offered her a taste. Winfrey gamely took a sip right on her show and declared the so-called "weasel coffee" eminently fit to drink. (Barry, for the record, opines that it "tastes like somebody washed a dead cat in it." Take your pick.)

What's the story here? The Asian palm civet, or luwak, is a small nocturnal mammal that's native to Indonesia, southern India, East Africa, Southeast Asia, the Philippines, and the south coast of China. By all accounts, luwaks are particularly fond of ripe coffee fruit. They digest the flesh of the fruit and excrete the beans, which are then gathered by grateful coffee farmers. So what's the attraction? According to one theory, the acids in the luwak's stomach dissolve the proteins in the coatings of the beans that cause the bitter aftertaste that accompanies more traditional blends. When brewed, kopi luwak supposedly has a mellower and sweeter flavor than regular java.

This theory has been put to the test by researchers at Canada's University of Guelph. In 2004, food science professor Massimo Marcone concluded that lactic acid bacteria in the luwak's digestive tract do indeed leach some of the proteins from the beans' outer shells. (It should be noted that most people in blind taste tests conducted by the same researchers could not tell the difference between kopi luwak and other coffees.) Marcone, who collected his own beans in Ethiopia and Indonesia, allayed fears of contamination by pointing out that coffee producers in Asia wash the luwak-derived beans extensively in order to banish any lingering bacteria.

No one knows who first decided to clean and roast the luwak's droppings. Kopi luwak was cherished in Asia long before Western importers decided to capitalize on its rarity and unusual origins. And capitalize they do: A single cup in one of the few American bistros that serves it will cost significantly more than normal joe. Who would have thought that a pile of poop could be transformed into a pot of gold?

Does a two-hump camel store more water than a one-hump camel?

First, let's give the poor beasts their proper names. A two-hump camel is called a Bactrian camel and comes from the plains of Asia. A one-hump camel is the most commonly known to Westerners; it's called a dromedary camel, and it hails from the Middle East and Africa.

Okay then, which one stores more water in its hump(s)? Um,

it's actually neither, despite what your teachers may have told you. What camels actually store in there is good old fat. That's right, the camel's hump is a bit like our beer belly, except the hump is far more efficient and useful.

The hump doesn't have much to do with water conservation. It's there to provide the camel with nourishment when it has no food. If the camel uses up all the fat in its hump, the empty hump will droop until the animal gets more food. (That's why malnourished camels often have droopy humps.) Camels can go for several weeks without food due to their humps—the bigger its hump, the longer a camel can go without food.

So the humps have got food covered, but what about water? We can't escape the fact that camels sometimes go days without water as well as food. Luckily, camels are really good at conserving water. When they have access to water, they drink between twenty and thirty gallons in about ten minutes—they can drink forty-five gallons in twenty-four hours. This water is then stored in their blood cells for later use.

Everything about the camel is designed so that it uses as little water as possible. It can withstand significant body temperature changes without needing to perspire; in fact, a camel can go from about 93 degrees to 107 degrees Fahrenheit without breaking a sweat. (If a human being's temperature goes up by two degrees, it indicates illness.) A camel's excretions are even very low in water.

So, does a two-hump camel store more *fat* than a one-hump camel? Yes, but the two animals are designed in such ways that the overall efficiency of the fatty-hump system is probably about the same on both.

What makes a firefly's butt glow?

Fireflies are nature's night pilots, diving and swooping like tiny prop planes, communicating their location and intent with a series of flashing lights. To potential predators, these lights say, "Stay away." To potential mates, they say, "Come hither." To a child running around the yard with a Mason jar, they can mean a lamp that will grow brighter and more fascinating with each bug that is caught.

Like all animals possessing bioluminescent traits, fireflies produce their light by means of chemical reaction. The bugs manufacture a chemical known as luciferin. Through a reaction powered by adenosine triphosphate and an enzyme called luciferase, luciferin is transformed into oxyluciferin. In the firefly, this reaction takes place in the abdomen, an area perforated by tubes that allow oxygen to enter, feed the chemical reaction, and become bonded to the luciferin and luciferase produced by the insect. Oxyluciferin is a chemical that contains charged electrons; these electrons release their charge immediately, and the product of this release is light.

All fireflies emit light in the larvae stage. This is thought to be a warning sign to animals looking for snacks: Chemicals in the firefly's (and the firefly larvae's) body have a bitter taste that is undesirable to predators. Studies have shown that laboratory mice quickly learn to associate the bioluminescent glow with a bad taste, and they avoid food that radiates this light.

In adult fireflies, bioluminescence has a second purpose: Some species of firefly use their glow—and distinctive patterns made by dipping and swooping, in which they draw simple patterns against the black of night—to attract a mate. Each firefly species, of which there are 1,900 worldwide, has its own pattern. Male fireflies flit about and show off, while the females sit in a tree or in the grass. The females will not give off light until they see a male displaying the wattage and sprightliness they're looking for in an attractive mate.

They must choose carefully, however, as there are certain adult firefly species that are unable to manufacture luciferin on their own. These species obtain the chemical by attracting unwitting members of other species and consuming them. They do this because without the chemical, they appear to predators much as any other night-flying insects; they need the chemical to advertise their bitter taste.

In the end, this effort might be for naught. The flavor particular to the firefly has that special something that some frogs cannot get enough of. For these frogs, the blinking lights are not so much a warning as a sign reading, "Come and get it!"

Who decides which bee gets to be queen?

Unlike the human business world, where grunts toil at their desks while the CEO zips in and out of his big corner office and hardly notices his minions, workers are the ones who hold the cards in a bee colony. In fact, they decide which bee gets to be queen.

A queen bee typically lives five to seven years. When she begins laying fewer eggs or becomes diseased, worker bees decide her reign has run its course. No golden parachute is provided, just a process called supersedure.

What's involved in supersedure? Workers quickly build a peanut-shaped queen cell, and a larva is raised in it. While this larva is identical to what develops into a worker bee, another average Joe isn't on the way. The larva for the queen is fed large amounts of a protein-rich secretion called royal jelly, which comes from glands on the heads of young workers.

After about two weeks, the new queen emerges. Just as two bosses are not good for business in the human world, the new queen seeks out the old queen and stings her to death. (Unlike the stinger of a proletariat bee, the queen's is not barbed and will not detach from her body upon contact with her victim. This means she can sting repeatedly without dying.) Reproduction is the key role for the queen because female worker bees are sterile. About a week after doing in her predecessor, the new queen takes her nuptial flights, mating high in the air with male drones. Then she begins her job of laying eggs.

Sometimes, though, things don't unfold as planned for the queen-to-be. Just as bosses in the human world don't like surrendering the big corner office, the established queen doesn't always go easily. There are occasions when the reigning queen defeats the aspiring one. When this happens, everything remains as it was in the hive—until another young challenger comes along and attempts to slay the queen.

The bees themselves don't always dictate the fate of a bee colony. Beekeepers, whose job it is to harvest honey, can initiate supersedure in the colonies they maintain. When beekeepers notice that the queen is laying fewer eggs, they clip off one of her middle or posterior legs, which prevents her from properly placing her eggs at the bottom of the cell. Workers detect this deficiency, and the queen is eventually killed off.

Would cockroaches really survive a nuclear war?

We've all heard that cockroaches would be the only creatures to survive a nuclear war. But unless being exceptionally gross is a prerequisite for withstanding such an event, are cockroaches really that resilient?

They are indeed. For one thing, they've spent millions of years surviving every calamity the earth could throw at them. Fossil records indicate that the cockroach is at least three hundred million years old. That means cockroaches survived unscathed whatever event wiped out the dinosaurs, be it an ice age or a giant meteor's collision with Earth.

The cockroach's chief advantage—at least where nuclear annihilation is concerned—is the amount of radiation it can safely absorb. During the Cold War, a number of researchers performed tests on how much radiation various organisms could withstand before dying. Humans, as you might imagine, tapped out fairly early. Five hundred Radiation Absorbed Doses (or rads, the accepted measurement for radiation exposure) are fatal to humans. Cockroaches, on the other hand, scored exceptionally well, withstanding up to 6,400 rads.

Such hardiness doesn't mean that cockroaches will be the sole rulers of the planet if nuclear war breaks out. The parasitoid wasp can take more than 100,000 rads and still sting the hell out of you. Some forms of bacteria can shrug off more than one million rads and keep doing whatever it is that bacteria do. Clearly, the cockroach would have neighbors.

Not all cockroaches would survive, anyway—definitely not the ones that lived within two miles of the blast's ground zero. Regardless of the amount of radiation a creature could withstand, the intense heat from the detonation would liquefy it. Still, the entire cockroach race wouldn't be living at or near ground zero—so, yes, at least some would likely survive. But they'd still be cockroaches. And how much fun would that be?